Kanban for your team

Tomek Kaczanowski

Kanban for your team

Tomek Kaczanowski

Website: http://kaczanowscy.pl/books

Published by Tomasz Kaczanowski kaczanowscy.pl, printed by CreateSpace createspace.com

Cover design: Agata Wajer-Gądecka, https://www.coroflot.com/agata1

Illustrations: Mateusz Popek, https://www.behance.net/user/?username=mateuszpopek

Translation from Polish: Paweł Nowak http://eklektika.pl

ISBN: 978-83-951851-0-6

First print: September 2018

Version: print_20180925_1814

Table of Contents

Chapter 1. Introduction

If you are looking for ways to help your team work more efficiently, this book is for you. Are you a team leader, or a manager? Or perhaps a dedicated programmer who feels that you and your teammates should be able to move mountains, but somehow that just doesn't happen? If you care about the future of your team, you should read this book. It's not even 100 pages long. You aren't risking much by reading it. At most, you will lose one hour of your life. And the potential gains are big!

This is not an exhaustive Kanban compendium. It is a practical and pragmatic guide, where I only describe what makes sense in the course of the work of a programming team, and, in particular, what proved to be effective in the teams in which I worked.

I am not claiming that the ideas presented here will be the perfect solution to the problems faced by your team. Perhaps, for reasons beyond your control, you will not be able to implement some of them. Too bad. If that's the case, try to find the gist of the techniques presented here, and find other means that can lead you to similar results.

Electronic version

The book is also available in electronic version. Details can be found on http://kaczanowscy.pl/books

Contact

I am open to (some) comments, to (constructive) criticism, to kindhearted advice (I'll bear with it somehow) and praise (in any amount!). So, if while reading this book you decide to give me some feedback of this type, please write to kaczanowski.tomek@gmail.com. Thank you!

Thank you!

When writing the Polish version of this book, I had the enormous support of many individuals. I would like to thank in particular those of them who read the book as it was being written and commented on it. Listed alphabetically, they are:

Paweł Dolega, Witek Graca, Szymon V. Gołębiewski, Mateusz Haligowski, Piotr Jachimczak, Wojtek Krawczyk, Kacper Kuczek, Jakub Marchwicki, Jan Mitkowski, Marek Rogoziński, Michał Rumanek, Marcin Stefaniuk, Maciek Wachowiec

Although many individuals supported my writing effort, and the final shape of the book is a result of the combination of my ideas with their comments, I hereby declare that all the imperfections are my fault alone!

I also have special thanks for colleagues from the teams where I had the opportunity to implement Kanban. I am very grateful for their trust, support, patience and constructive criticism.

Icons

The following icons appear in the text:

A helpful tip

Additional information

Be careful, danger is near!

All the three icons were designed by Webdesigner Depot http://www.webdesignerdepot.com/.

Chapter 2. Dictionary

"When I use a word," Humpty Dumpty said in rather a scornful tone, *"it means just what I choose it to mean — neither more nor less."*
— Lewis Carroll *Through the Looking Glass*

This book is largely based on the experiences of programming teams, which is why it includes terms that are characteristic for the IT industry. I've decided to explain them here for those readers who have not had a chance to learn them. It is not necessary to understand these terms to comprehend the contents of this book. Having said that, the book may be a much better read if you get acquainted with them beforehand.

- **Backlog** - the list of tasks to do. In the Scrum terminology, it may apply to all the tasks (*product backlog*) or to tasks chosen for implementation in the given Sprint (*Sprint backlog*).

- **Bug tracker, Issue tracker** - an electronic system for the recording of errors or tasks. For example, JIRA, Bugzilla or Trac.

- **Continuous deployment** - a way of organizing the work of a developer team where everything that gets to the code repository hits production just moments later. It requires high code quality (or tons of luck!)

- **Continuous integration** - combining the code written by many programmers into one whole. Its aim is to avoid long periods when individual team members work on their own. After a while, many problems arise from the integration of code created in this way. In the case of continuous integration, each team member puts code in the common repository at least once per day.

- **Continuous integration server** - a continuous integration tool, responsible for building projects (primarily compilation and the launching of automated tests). Examples include Jenkins, Bamboo or TeamCity.

- **Commit** - the act of placing code in the code repository where other team members can retrieve it from, and where the continuous integration server retrieves it from as well.

- **Code review** - inspection of code aimed at discovering errors and propagating knowledge among team members.

- **Cross-functional team** - a group of people with diverse skills, selected in such a way that they can pursue goals without reaching for help from outside the team.

- **Deadline** - the date by which a task must be completed, or we can expect trouble: a financial fine, a reprimand from an Important Person, or some other, equally charming, niceties.

- **DevOp** - a team member who is responsible for deployment to production, for the configuration of test environments etc.

- **Definition of Done**, also known as **DoD** - the criteria for considering a task completed.

- **End-to-end tests** - tests in which the entire system is verified.

- **Feature** - something to be implemented. Typically, it's a new functionality.

- **Information radiators** - visual and acoustic solutions that ensure that some information reaches all the interested parties.

- **Integration tests** - tests that verify whether individual fragments of the system (e.g. modules, layers) cooperate correctly with each other.

- **Pair programming** - work in pairs. A coding technique where two developers cooperate on one objective (e.g. adding a new functionality to an application).

- **Project Owner** - the person who knows what we are doing, and why. In particular, the Project Owner knows which tasks are currently priorities.

- **QA** - *Quality & Assurance* – a person or a team of people who are responsible for ensuring quality. When the entire team is responsible for product quality, the role of QA is more about supporting developers in this respect rather than relieving them of testing duties altogether.

- **Scrum Master** - in a team that follows the Scrum methodology, the Scrum Master is a person responsible for the smooth progress of the team's work, by ensuring that Scrum guidelines are correctly followed and that impediments that make it difficult for the team to complete tasks are removed.

- **Sprint** - a period of time during which the team (usually a Scrum team) undertakes to provide certain value to the client (which frequently entails implementing and deploying specific functionalities to production servers).

- **Unit tests** - tests which verify the correct functioning of the smallest code elements, in particular individual classes.

- **Waterfall** - a project management methodology which, as the name itself indicates, leads to a fall... Interestingly, it is typically attributed to the person who has been critical of this approach to project management from the very start.

Chapter 3. About Kanban

Kanban is **a set of rules and practices for the management and optimization of the production process**.

Kanban was born in the 1950s in Japan. It promised to make industrial production more efficient by eliminating unnecessary elements of the process, delays, downtimes, queues, idleness and stock building. It delivered. Thanks to the use of Kanban (and the implementation of related techniques such as *just-in-time*[1] and *lean manufacturing*[2]), many Japanese companies have greatly improved their results.

Kanban grew out of manufacturing. It was developed by Taiichi Ohno, a Toyota employee. It quickly turned out that Kanban can be applied to various types of endeavors and problems. Its use in software development can primarily be attributed to David J. Anderson, who authored the first implementations and books on this subject.

What fascinates me in Kanban is the ease with which it can be applied, and this is why this book focuses on the practical side, with less attention given to theory. However, before we move to practice, it is worth learning the basics of Kanban.

3.1. Principles

> Obey the principles without being bound by them.
> — Bruce Lee

These are the four Kanban principles[3]:

1. Start with an existing process.

   Kanban does not define any specific roles or process steps. Kanban is a factor that stimulates the evolution of an existing process and roles in a new direction. Kanban is a change management method.

2. Accept gradual, evolutionary changes.

---

[1]https://en.wikipedia.org/wiki/Just-in-time_manufacturing
[2]https://en.wikipedia.org/wiki/Lean_manufacturing
[3]Based on Wikipedia https://en.wikipedia.org/wiki/Kanban_%28development%29.

Big, sudden changes generate fear and opposition. The way to improve a system is to continuously introduce gradual, evolutionary and often small changes.

3. Respect the current process.

   It is not the goal of Kanban to destroy the current processes and principles. What already exists and works represents an accomplishment of the company and its employees. You should build on these accomplishments, taking the good stuff, improving the imperfect stuff and getting rid of the faulty stuff.

4. Everybody is a leader.

   Every team member can make a valuable contribution to process optimization. Initiative and self-reliance should be respected and supported.

3.2. Rules

> Never bend the rules. You bend the rules a little bit and then it's a slippery slope.
>
> — Thomas Peterffy

A Kanban team follows the principles discussed above by applying the following rules:

1. Visualize.

   The first step to improve the situation is understanding where we are now. Visualization makes it possible to see what the team's work looks like, and what the weak points are.

2. Minimize the number of tasks worked on.

   Each team has a specific capacity. The number of tasks worked on should be minimized in order to focus on completing the ones which are currently open.

3. Manage the process.

An effort should be made to make the software production process as fluid as possible. Collecting data on the elements of the process makes it possible to monitor individual stages of production.

4. Apply clear rules.

   Let the software production process be governed by clear, simple and transparent rules. This way, you can discuss it based on facts rather than guesses.

5. Make sure there is feedback.

   Feedback is necessary for the team to know whether the changes being implemented are for the better.

6. Streamline and experiment.

   Kanban reveals the true nature of the production process, making it easier to see the areas that can be streamlined. By using problem-solving methods, the team can plan changes, implement them and then verify their results.

In the ensuing chapters we will see how the principles and rules discussed above can be put into practice.

(i) A mere repetition of the typical Kanban activities, without making an effort to understand what motivates them, is unlikely to bring great results. It is a good idea to remember the Kanban principles, so that we can ensure that our actions are consistent with them.
Otherwise, we risk becoming cargo cult followers[4].

3.3. When to use?

There are many reasons to introduce Kanban practices into your team. You should take interest in it if you are facing one of the following problems:

• Team members do not know which tasks are important. When they complete a task, they frequently choose a low-priority one as the next one.

---

[4]See https://en.wikipedia.org/wiki/Cargo_cult

- The "top brass" are annoyed that the team has done "only this much" in the past days or weeks. Team members, on the other hand, feel that they have been working a lot, and hard.

- Project manager and/or owner does not know what the team members are currently working on.

- Scrum "does not work". The team does not manage to complete the task to which it has committed in the given Sprint. Frequent "surprises" in the form of new tasks, resulting from mistakes or changes to task priority, disorganize the planned work.

- In spite of daily meetings, team members do not know what their colleagues are doing. This sometimes leads to mistakes with disagreeable consequences: *"I thought that had been in production for a while now!"*

- Long time estimation sessions are tiring for the team, and the resulting estimates are frequently worthless.

What can we expect?

As a result of introducing Kanban, we can expect positive changes in the team's work. All the team members will have improved visibility of what is currently happening in the project. Just one look will be enough to find out which tasks are currently being pursued, which ones have been completed and whether *"feature X is already in production"*. Team members will receive a huge energy and enthusiasm boost in the form of a tool that clearly visualizes their everyday efforts. Managers and team leaders will find it easier to localize bottlenecks, and thus to reorganize work to improve efficiency.

This sounds really good. And it's just the beginning of the positive changes that the team will be seeing as Kanban is used.

Chapter 4. Just do it

The best time to start was yesterday,
the next best time is **now**.

— Common knowledge

Kanban is so simple that learning it is best started by… starting to use it!
This is what we will turn to now. We will jump into the deep end of the
pool, and I promise that it will soon turn out not to be so deep after all!

We will then clarify some things, but, as is usually the case in life, you will
benefit the most by analyzing what your team can (or cannot) do. So, let's
start our work with Kanban and let's draw conclusions from the progress.
In the ensuing parts of the book, some important topics will be discussed
in more depth, but there's nothing quite like first-hand experience.

Let's go then!

A few things will be needed:

- a board – a dry erase whiteboard or a corkboard[1],

- sheets of paper,

- a solid marker,

- magnets or pins (depending on the type of your board),

- support of your team members (or at least non-obstruction on their
  part…).

The first four prompts can be ordered from the appropriate department of
your company, while the fifth one will require some work on your part.

---

[1]We will talk about electronic boards in 7.1. Until then, let's not get distracted.

4.1. The board

For starters, let's think about the phases into which we can divide the typical tasks that team members perform daily. As the starting point, let's take this diagram, which is applicable to the work of essentially any team:

TO DO	DOING	DONE

Just three phases:

- *ToDo* – what is to be done,

- *Doing* – what is happening now,

- *Done* – what has been finished.

It may be a good idea to expand this diagram at the very start, so that is better captures the phases of the work on development tasks. The boards of most of the teams I know have a *Test* column between *Doing* and *Done* for the tasks which require such verification.

TO DO	DOING	TEST	DONE

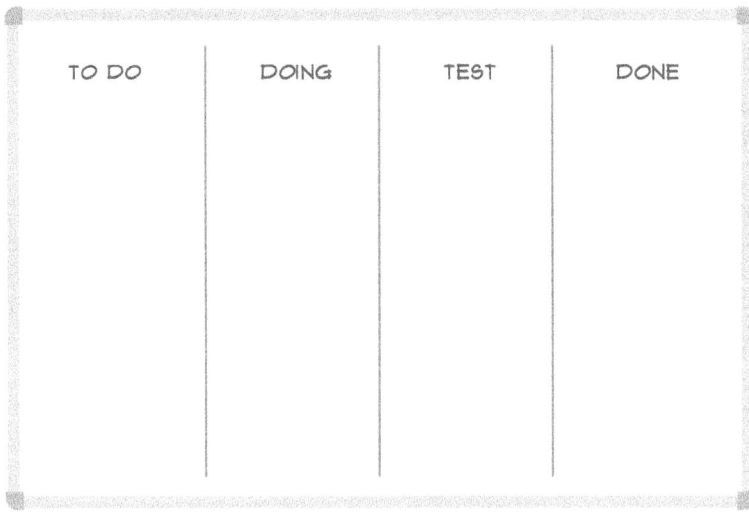

I will not try to impose a single correct approach, because it does not exist: the columns on the board depend on the specifics of your team's work.

Reality rather than dreams

The Kanban board should reflect the current working process. The way it is now. It is not a wishing well nor the reflection of a better tomorrow. So, even if you are not fully happy about the way your team is working now, make sure it is shown faithfully on the board. Perhaps you would like each task to be preceded by an analysis stage. If your team does not do this, however, do not put an *Analysis* column on the board. Put a *ToDo* column followed immediately by a *Doing* column if that's what happens.

By analogy, let the columns on the board show all the stages of the production process that are important (i.e. they take time and engage people). If tasks go through code review, the board should include such a column. If, in order to be allowed into the test environment, they must be approved by the Grand Architect (let his beard be luxuriant and his UML diagrams even more luxuriant), then add a column to the board to represent His Highness. And so on.

On the other hand, it is not the case that the system of columns that you draw legitimizes the current process. No way. The board only reflects reality. It is a visualization that is meant to make some things very clear. Neither the current process, nor the board are sacrosanct. As Kanban is used, the team's rhythm of work will change, and this will eventually be reflected on the board, for example in the form of new columns, or in the removal of existing ones.

4.2. People

Let's assume that you've gone through the production process, and based on this you've planned the columns that you are going to put on the board. You have also ordered all the necessary components from the appropriate department. There is one task remaining: to win the team members' acceptance.

A lot depends on your position within the team, the relations between colleagues, the type of people you are working with, the current customs and the broadly understood working culture inside the team, department and company. I can just give you a couple of arguments that you could employ, leaving to you the decision about which one to use and how when talking to your colleagues.

My experience shows that developers are generally in favor of ideas that streamline the team's work, especially if they are aware of the drawbacks of the current approach. So, it could be the case that some members of the team will be in favor as soon as you mention the idea. It may turn out that some of the team members have heard about Kanban, while others will simply be happy to try something new.

My experience shows that there are three basic causes of resistance to Kanban:

- First, team members may be tired with the implementation of a new methodology (*"another genius theory, and the effects will be the same as ever"*). In such a case, all you can do is to ask them to give you the benefit of the doubt so that they can see that this time is different. You will not convince extreme defeatists anyway (*"this won't help, it won't solve our problems, all this is vanity of vanities"*); you can just ask them to quietly await the tragic end without stirring problems.

- Another reason is fear of additional work due to the need to take care of the board and to create the cards. Here you can allay their fears by saying that this additional work is truly negligible (which is true).

- The third reason for resistance is a certain feeling of threat resulting from increased visibility of team members' work. Let's be honest:

there are developers who do not like to be observed while working. They would be the happiest working under a manager who won't challenge them when they spend three weeks on getting acquainted with the nuances of technology X. Kanban will significantly increase the visibility of all actions (we will discuss that at length later). Most likely, such persons will not protest when Kanban is being implemented. Some time will pass until you notice that not everybody is happy with it. It is also very likely that you will never be explicitly told what's the reason for these unfriendly looks directed at the board (and at you!) But if you can read between the lines what they are saying (*"why make these cards if I say during every standup what I am doing, don't I?!"*), you will figure out what the problem is. You've rubbed them the wrong way, you've limited their liberty for the sake of the project and the team. Gosh, what are you going to do now?

It is not my goal to scare you about the potential resistance of the team. My experience shows that people are generally quite favorable towards new concepts. It is worth knowing, however, what you can encounter. You should therefore prepare well for the discussion.

You can try convince the skeptics by showing them the benefits of Kanban, in particular the improvements in communication (both among developers as well as between the team and the management), the minimization of the number of tasks performed in parallel, the removal of downtimes and the opportunity to constantly improve the organization of work. You should emphasize the elements that are the team's weak points now. This should give you enough support to continue working.

Let the Power be with you!

4.3. Where do we stand?

> If it's a unit of work, visualise it, so we have an accurate view of the world.
>
> — Christ Roberts

Let's say the prompts are ready and the team is prepared to experiment with Kanban. Great! This means it's the time for a meeting of the entire team. **Entire**, that is including project owner and/or manager.

The goal of the meeting is to agree on what the current state of the work is. This requires writing down all the tasks that are currently being pursued. Each of them should be written on a separate card which should then be put in the appropriate column of the board.

(i) This is a bit like the GTD method (Getting Things Done)[2]. First you have to see how much you've got on your plate; only then can you start to *clean up*. Without a sincere description of the current state, an improvement is impossible.

On the board shown below, different types of tasks are marked with differently colored cards (we'll discuss this a bit later). You can start by using identically colored cards.

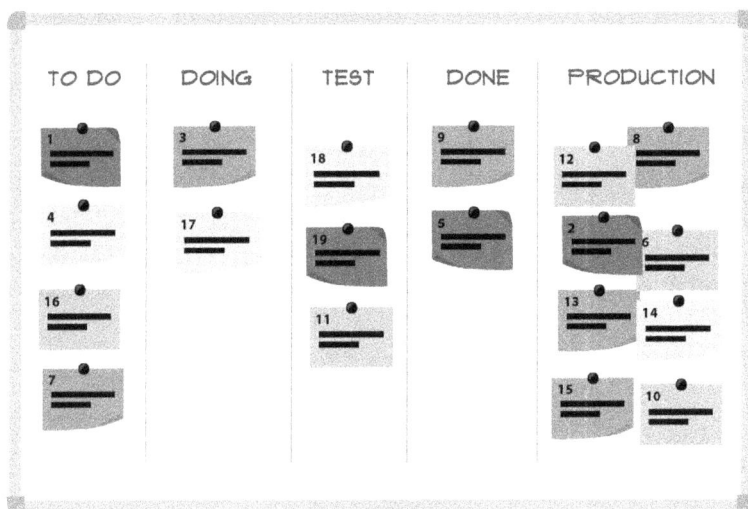

At this stage, the key point is filling the columns that symbolize the stages of the software production process during which *"something really happens"*. Usually, this is a simple thing, since every team member is capable of describing what he or she is doing.

The first column, i.e. *ToDo*, may be a challenge of sorts. We can start by assuming that it should include the tasks we expect to start working on

---

[2]http://gettingthingsdone.com/

within a week. We will introduce many modifications in this column (for sure during team meetings, and perhaps on daily basis), so there is no need to dwell on it much right now.

The final column, which contains completed tasks, will be empty at first. No problem. It will get filled soon.

> (i) One task, one card. No matter how many persons are working on it.

4.4. Cards

We have said that all the tasks should be attached to the board as cards, but we have not discussed at all what such a card should look like. Let's look at a sample card then.

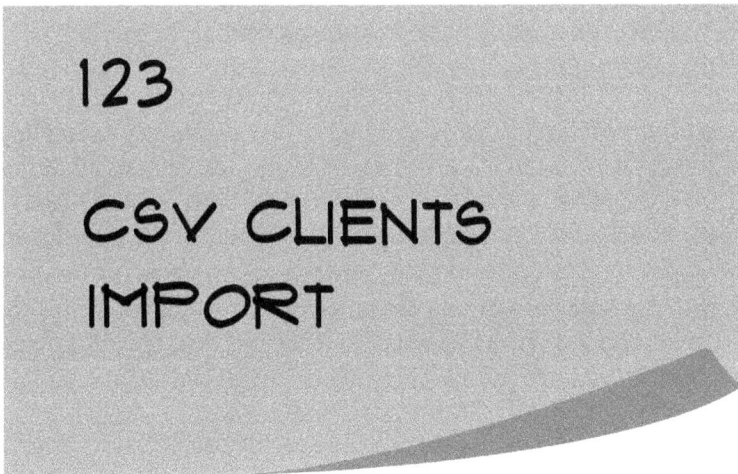

Not very complicated, right? Right. And let's keep it that way.

The card contains just two pieces of information. The first one is a short description of the task. The key word is short. So short that it can be written in a large font and that it can be understood at a glance. No complex sentences, no detailed descriptions.

The details belong in the issue tracker. This is where you describe the task in detail, include an outline of the architecture, paste stack trace of a production error, link other tasks etc. There is no room for that on the card.

There is room, on the other hand, for issue tracker task number. Those that want to do so can click through and read the details[3].

If the cards are illegible, the board will not be easy to understand and it will not communicate valuable information at a glance. And if something is not readily available and requires an effort, we simply do not use it.

There should be cruel and unusual punishment for those who scribble illegibly on the cards. For example, you could ask them to write database triggers, or to optimize the presentation layer for IE6. More seriously, you should make sure that cards are legible, asking the authors of the illegible ones to rewrite them. At some point they'll learn.

4.5. First effects

You can observe a lot by just watching.
— Yogi Berra

Perhaps already at this stage you've learned something interesting about the project that you thought you knew inside out. First of all, it may have turned out that team members perform more tasks than the issue tracker suggests. Someone is preparing and conducting an internal training. Someone else, from time to time, "just has to" implement a fix for another team, in the code he used to work on. Someone prepares a biweekly report specially requested by a manager from another department. Someone has managed to lead two job interviews this week. And so on.

There's a good chance that seeing the true picture of the work performed in the project, the project owner (and/or manager) will be appalled. And the team members themselves can be surprised by the number and the variety of the tasks.

Don't shoot the messenger! Visualization has merely revealed the problems that have been troubling the team.

It's not just knowing how many tasks there are in the project that can surprise us. The fact that everybody can see what other team members are

---

[3]More on that in chapter 7.1.

doing sometimes leads to huge surprises. For example, someone discovers that a colleague is working on task X, much to his surprise, since task X was completed on his local machine six months ago – but, for reasons that are no longer relevant, the code just didn't make it into the repository. Someone else may be indignant at not seeing on the board the tasks that, in their view, should be performed now.

(i) To hear something during a daily meeting, and to see it on the board, are two very different things. We see more, and what we see makes a stronger impression on us.

4.6. Ready? Go!

We know our starting point, so it's the time to start! It's time to roll up our sleeves and get to work. It's just a day like any other – everyone has a task they must face. With one difference from the way work used to be done. Starting today, every action taken must end up on the board. As the work progresses, team members will move the cards that represent their tasks to the next column. And when new tasks arrive, they will be put on the board as well.

Notice that **nothing has really changed in the way the team works**. No revolution, no deep transformation. Everyone continues doing what they have been doing until now. Like writing code, testing, retrieving lost data from production and recording the progress of the work in the issue tracker. The code still makes it into the repository, and the continuous integration server keeps building it after every commit.

That's one of the features of Kanban, which is **not a process** and does not attempt to replace an existing process. To use the programmers' language, I'd say that Kanban adds another abstraction layer to reveal what has been hidden.

It's quite likely (actually, nearly certain) that as the work with Kanban progresses, it will become clear which elements of the current process can be improved, but for the time being let's be content with what we have.

And thus, with practically no pain, we have entered a new world; a world whose center is the Kanban board. Of course, there are still many clarifications and additions ahead of us, but we've made the first step. More will follow.

4.7. Watch out for entropy!

Isn't that great? We have the board and the cards and everything is going smoothly. But can we be sure? Experience shows that **things don't happen by themselves**. If we do not ensure things are in order, disorder will soon reign.

We need someone to take care of everything. By making sure that pens and cards are available, and that each task makes it onto the board. By ensuring that the layout of the cards on the table always corresponds to the state of the work in the project. By notifying everybody about meeting times (to be discussed soon) and ensuring that they turn up for them. In other words, someone to take care of all the things related to Kanban.

In some teams, the Scrum Master does all this, in some others, the project manager, and finally in yet other ones someone with the strongest sense of responsibility for the team.

(i) Of course, if it was you, dear reader, who suggested Kanban to your team, you've got all the chances in the world to be privileged with the "process maintenance" task. Be prepared for this.

A lot depends on the team's maturity and on how convinced they are about using Kanban. In the optimistic version, it may turn out that such a "Kanban guard" will only be needed in the first few weeks, because afterwards all the team members will divide responsibilities among themselves. And if someone takes an outside look at such a team, they may conclude that things are, after all, happening all by themselves. Experience shows, however, that from time to time **someone** is needed (you?) who will remind the team of the rules concerning the placement of the cards on the board.

Chapter 5. Principles

> Do not plan for ventures before finishing what's at hand.
> — Euripides

A team that works in the Kanban spirit should always remember about the objective of the work, which is to **provide value to the client**. Therefore, completing tasks is the priority. Nobody gives points for starting another task. No client will be in awe of a task that is 'done', but has not been tested yet. The end client is interested in an error-free, working and complete functionality. And, which is worth remembering, they usually want it as soon as possible.

As one of our Kanban sayings goes: *"stop starting, start finishing"*. Let's think together what that means in practice.

Let's assume that the team's board looks as shown on the following diagram:

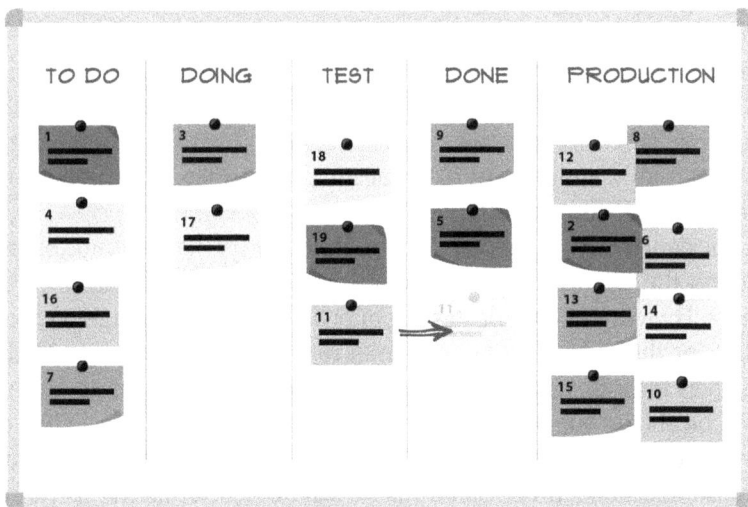

Now, let's imagine that the developer working on the task represented by card number 11 has finished working. So she moves it to the column marked *Done*[1]. Then, after a short break to celebrate and have coffee, she faces a dilemma that can be summarized as follows: *"what should I start*

---

[1]And probably updates the status in the issue tracker, or notifies the interested parties directly.

doing now?". The Kanban answer is: *"do what will bring value to the client"*.

Done

Let's look at the board. Only the tasks that are in the *Production* column bring value to the client. If so, the developer should work on implementing the stories that are in the *Done* column in production. Of course, there are as many rules as there are teams. Perhaps deployment is done once a month (hopefully not...), perhaps we are awaiting the green light from an Important Person, perhaps we've agreed with the client that the next deployment will include the (not yet completed) functionality Y. And so on. I am not saying that it's the duty of our developer to do the deployment. I am just pointing to the place where her actions will bring the biggest value to the client.

Is it possible in the given situation for the efforts to be directed there? I am not able to assess this. But deployment to production would certainly be good.

Test

Let's assume, however, that deployment to production is not possible at the given moment. We look further left and concentrate on the *Test* column. There are two tasks there. Since they are there, someone must be working on them already. The question is: can we join these efforts and help complete them earlier?

A lot depends on the context here. Frequently, that will not be possible due to the way the tests are organized, or due to the nature of the test tasks. By way of example, testing may be the domain of highly specialized staff, or the tasks may be too small to enable reasonable parallelization of their testing. However, sometimes it may turn out that our developer will be able to support her colleagues. Perhaps she will do a code review, while the QA division will focus on writing end-to-end tests. In a word, she should make an effort to find a way to accelerate the work on tasks in the *Test* column.

Doing

However, if that is not possible either, the developer should look for work in the *Doing* column. There's a big chance of finding something here. Many nontrivial tasks can be divided into smaller ones, which could make doing them faster. Another positive consequence of dividing the tasks is propagating the knowledge of a given functionality among more team members.

Another way of getting involved in work is pairing up with someone. As the saying goes, two heads are better than one, and everybody who has ever done pair programming knows that this is still true.

ToDo

It is only when it turns out that there is nothing to do in the *Doing* column, the developer should look for new tasks in the *ToDo* column. Here, in turn, the question might be which task to pick (the column will usually have several ones). Well, they are either prioritized, which makes decisions easier, or the situation should be discussed with other team members. Of course, we also need to take into account the given team member's skills, and their knowledge of the individual areas.

Let's finish!

Let me repeat here the following: **the only stuff that has value for the client is what has been deployed to production**. This is obvious, but many developers are still reluctant to work on tasks in the *Test* column while so many new, tempting adventures are awaiting in the *ToDo* column!

I do not know a universal prescription how to make team members behave in line with the ideal described here. Personally, I attempt to reward (with a good word) all the examples of such behavior, and I encourage team members by reminding them how important task completion is.

5.1. Kanban is a way of thinking

As our example shows, a developer who acts in the Kanban spirit looks at the board from right to left in search for the value that could be given to the client. This requires:

- communication with other team members (including project owner),

- joint work on tasks by several people,

- going beyond one's *comfort zone*,

- thinking in terms of the value delivered to the client.

(i) (Non-)accidentally, such behavior is exactly what we would like to see from all team members!

As an additional commentary to these considerations, let's add a few words on Kanban. Using it is supposed to result in smooth, unobstructed flow of tasks. The key point is finishing tasks while making sure that at no point too many tasks are open simultaneously.

The columns on the Kanban board do not delineate the responsibilities of individuals! There's no room here for the so-called *over the wall attitude*, which means that specific groups keep passing tasks to each other on the principle that *"we've done our work, it's your problem now"*. The behavior that we know from production processes based on the infamous waterfall are not allowed here. The columns on the board are just phases of the smooth transition from the beginning to the end. They exist to make it possible to visualize the progress of the work and to discover bottlenecks.

5.2. Unobstructed flow

The ideal situation is the continuously flowing stream of tasks. The romantic souls will think about water flowing in a peaceful river with no obstruction. Those more engineering-minded might think of an automatized assembly line which delivers completed products one by one. Both of these images are correct and they both represent well the gist of what we are striving for. We are interested in **constant, unobstructed flow**.

Of course, nothing is as simple as it seems. The key to avoid situations when our assembly line gets stuck is maintaining high quality. If we can't do this, the smooth flow will be disrupted. Instead of moving steadily from *Doing > Test > Done > Production*, the cards will be returning from *Test* to *Doing*. This, in turn, will cause developers to switch context, ditching current tasks and moving to error fixing.

It's even worse when testing is also sub-standard. Then, the team will attain a temporary success (i.e. deployment to production), which will be followed by a catastrophe when the task is tested by a client. And, since a lot of time may have elapsed since the work ended on the given task, fixing it may turn out to be much more costly. The resulting disruption of the flow will be even more clearly visible.

Another important aspect that supports the smooth progress of work is lack of dependence on non-team-members. Being dependent on other teams, e.g. database specialists, makes the team lose control over the pace at which tasks are flowing. It is hard to push a blocked task forward while waiting for actions by third parties. Unobstructed flow becomes just a dream. What remains is begging, arguing or bribing the members of other teams to prioritize our tasks.

In short: It is good to be independent. It isn't therefore a bad idea to build a cross-functional team.

> ## Cross-functional team
>
> A cross-functional team is a team that has all that is needed to complete its tasks. For example, there is no need to ask the QA team for help, because there are such people on the team, or team members have such skills.
> Let's be honest – something like this is extremely rare. Companies typically have employees with unique skills (e.g. UX designers, content writers, super-hyper QA masters) whose services are used by various teams. Which does not alter the fact that it is a good idea to have people with varied skills on the team, and to have to reach outside only occasionally.

5.3. Bottlenecks

We have an intuitive understanding of what a *bottleneck* is: It's the stage in the process when the progress of the work is at its slowest. This is where tasks are accumulated in such a way that people involved in the further stages of the process must wait (idly) until the bottleneck lets through the next element ready for further processing.
An example of a bottleneck is a single cash register in a store with a long line in front of it. For a development team, the bottleneck could be, for example, the analysis stage, with few people involved in it. What could happen is that programmers will have to wait for tasks on which they could work.

Let's notice that in both cases the bottleneck determines the bandwidth of the entire system. The whole will not become more productive until this very element is made more passable. In the case of the store, making the aisles between shelves wider will not help. In order to increase the bandwidth of the store, it is necessary to either accelerate customer service at the cash register, or add more cash registers. In the case of a team that produces software, hiring better programmers won't do. Only changes to the analyst team can improve the situation.

> (i) A bottleneck can happen at any stage. Please, treat the task analysis deliberations below as examples. We could just as well consider task accumulation in the *Doing*, *Test* or *Done* columns.

The best solution would be to improve the efficiency of the element that constitutes the bottleneck of the process. Above all, it is worth making sure that the individuals who are meant to perform the analysis are not burdened with other tasks. It's quite possible that that's where the problem lies. The analysts may be involved in more than one project, which limits their progress on our process.

If we do not have the power to *take the analysts over*, we should try to make them more available for us. Let's assume that there is only one analyst available, who works for us on Wednesdays and Thursdays. If we can change his schedule so that he turns up on, say Mondays and Thursdays, we will see some improvement. It's quite possible that on each of these days he will be able to analyze a task. In that case, subsequent "portions of work" will be reaching us regularly, and not just once a week. This will lead to a more steady pace of work for the entire team.

Next, we can try moving people between positions in order to fix the situation quickly. In the case of the problem with analysis that we are discussing here, this would mean transferring some programmers to task analysis. When such a boosted analysis team has coped with the accumulation of tasks, the programmers could return to their usual tasks. There are some risks here, however.

First, it is possible that employees who are not skilled in analysis would not do it as well as experienced analysts[2]. We are therefore

---

[2]Of course, in the long term such a move could prove to be very valuable, especially if the analysis and coding divisions have been strictly separated until now.

risking a reduction in analysis quality, and that could prove to be very disadvantageous.

Second, switching context is deleterious, and delays in the implementation of programming tasks could be difficult to accept. Programmers will be first forced to adjust to completely new tasks, and then to return to their usual ones.

What is more, by transferring people there and back, we will not eliminate the root cause, i.e. inefficiency of the analysis stage, and every now and then we will be forced to repeat the same maneuver. Paying a hefty price for that each time.

Looking at the problem in a longer perspective, one thing seems certain: It makes no sense to bury the analysts with tasks in the amount that exceeds their bandwidth. Until we have strengthened the analysis department, we should not expect the overall system to have a higher bandwidth than that department does. Thus, it seems reasonable to limit the number of tasks performed in order to ensure smooth, unobstructed task flow.

(i) The Kanban board is an excellent tool for discovering where the bottlenecks are. The accumulation of cards at the entry into the bottleneck, with few cards on the other side, is a clear signal that this is where streamlining efforts should be focused.

Chapter 6. Work in Progress

> Don't spread yourself too thin.
>
> — Common knowledge

WIP, or *Work in Progress*[1] is the number of tasks that the team is currently working on.

All the tasks which have been started, but have not been completed yet, increase the WIP. Thus, if the first column on our board is *ToDo*, and the last one is *Production*, then WIP includes all the tasks that have left *ToDo*, but have not been deployed to production yet. It does not matter whether we are actively working on them, or whether they are waiting for the QA team to have sufficient bandwidth to take care of them. This is not important. What matters is that the work has been started, that they are "hanging over our heads", that they cannot be crossed out and forgotten.

Unclosed loops

Your question could be what the cost of tasks which have not been completed and are still "hanging over us" is. None, in theory, quite big in practice.

These are like the unclosed loops from the Getting Things Done. They do not seem like much, but in an imperceptible way they are taking away some of our bandwidth. We are not doing anything with them, but we are still thinking about them and getting frustrated when, time after time, we realize that this task is still not completed.

In the case of development, the cost of such unfinished tasks may even be higher because they usually have to do with unfinished code, which lies somewhere in the repository, perhaps making work on the next tasks more difficult.

6.1. Penalty for high WIP

Thus, each open task leads to some additional work and to a number of problems. Let's have a look at them.

---

[1] or *Work in Process*.

- **Context switching**. With many tasks pursued in parallel, we are forced to switch between them. Every time we do this, we are losing time. That's just the way we are – "entering" a new topic fully can take as much as 15 minutes. In addition, we have to record (in memory or in writing) where we interrupted the previous task.
  Every now and then we will be making mistakes: we might think that we have already done some things (when in fact we were just thinking about them), or that some things have only been planned (while we have actually done them). In both cases, our work will lack a steady rhythm.

- **Re-discovering knowledge**. Each time we are returning to an open task, we must understand its nuances again. Depending on how complex the task is, and the time that has passed since we last worked on it, this could take a moment, or much longer. The real problems begin when we are continuing a task that was started by another person...

- ... because that's where we are hitting the **knowledge transfer** problem. Experience shows that some things are not transferable, and no type of documentation can really help with that. If we are lucky we might get a chance to talk to our predecessor, but even that will not allow us to get all the knowledge. And we will have to reinvent something that has already been invented.

- The more tasks being done in parallel, **the more delayed their completion will be**. And we will deliver them to the client later. And this results in:

- **Delayed feedback**. The later we present a demo or deploy to production, the later we will learn what the client thinks about a given fragment of the system under construction. And, in the case the client is not fully satisfied, we will pay a higher penalty than we would have to pay if we had allowed the client to review our work earlier. There will be more to change, redo, and rethink.

- As the development time increases, the risk increases that changing business requirements will **reduce the importance of the task for the client** (vs. earlier delivery). In extreme cases - when a deadline is exceeded - the task may lose all its value altogether.

(i) Imagine a dual carriageway with five cars. All of them are driving quite fast and will reach the target soon. Now imagine there is fifty of them. The resulting traffic jam will make the journey time much longer for all of them. This is exactly the same with the team of programmers – the more tasks are being worked on, the longer it will take to complete them.

6.2. Morale losses

Until now, we were thinking about the impact of the high WIP on project completion. It seems that delivering it on time may be at risk when the team's efforts are spread onto too many parallel tasks. Let's now have a look at the impact of WIP on the team members' morale.

Let's start with a short quote from the personal development guru Brian Tracy:

> Each time we complete a task we get an influx of energy and enthusiasm, and our self-esteem increases considerably.
>
> — Brian Tracy *Eat that frog*

When WIP is high, every team member is occupied with more than one task during the day. And this means they will complete only a portion of each. So you can forget about the enthusiasm influx related to task completion.

Of course, such a situation is not without impact on the team member's mental states. Working on one task, you are thinking of another (which is also urgent and important), which leads to stress. When several things are being done at once, usually it is not possible to complete any of them. Hence the feeling of lack of achievement at the end of a working day: so much work but nothing really done. Add to this the constant context switching and having to remember things related to both the first and the second task (and perhaps also the third one and the fourth one). It is hard not to feel overwhelmed and not to get frustrated. None of these emotional states helps you work more productively...

What is more, the client is frustrated as well! And so is your boss. And his wife, because she is annoyed when he is getting home and he pays no attention to her new haircut. There's a storm brewing!

6.3. WIP limits

What we said above clearly shows that a large number of tasks open simultaneously is not good. High WIP leads to time losses and has a bad impact on the employees' mental states. No wonder than that one of the major goals of Kanban is *limiting WIP*.

One commonly used method of pressurizing the team to limit WIP is imposing restrictions on the number of tasks that can be put in the individual columns of the board.

These are usually marked on the board by putting a number in brackets next to the name of a given column. These numbers set the limits, i.e. the maximum number of cards that can be put in the given column.

Let's have a look at the board shown below. The following limits are marked there: 3 for the *Doing* and *Done* columns and 2 for the *Test* column. They determine the maximum allowed number of cards in these columns.

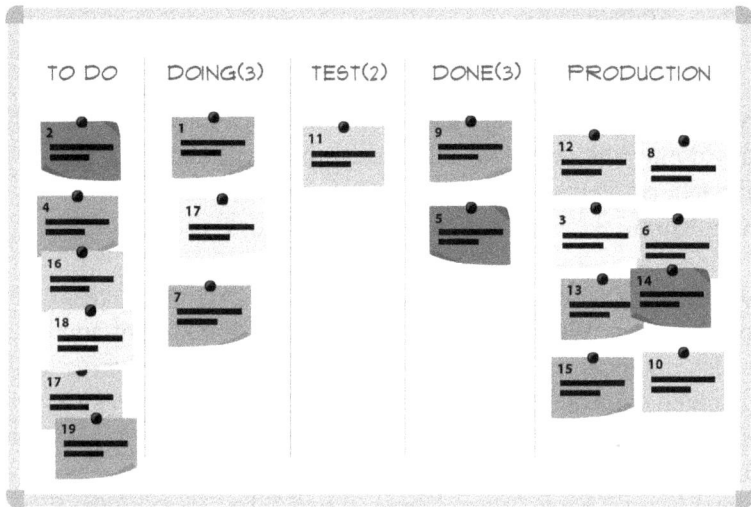

The setup of the cards on the board indicates that no card can be added to the *Doing* column (because it already has 3 cards, i.e. equal to the limit). So, what should a developer do while looking for a new task upon completing a previous one? He should think about working on one of the tasks that can already be found in the *Doing*, *Test* or *Done* columns.

Note that this exactly repeats the considerations that we were engaged in when deciding how to deliver value to the client as quickly as possible.

The difference is that at that time we were appealing to the developer's common sense, while here we have a clear delineation of his freedom. Sufficiently low WIP limits force team members to share work on tasks, to unblock blocked tasks and to work in pairs.

WIP limits are not meant to be a spiteful interference in work. They are meant to put pressure on the team to increase cooperation, to take responsibility and, in effect, to complete tasks earlier. They are meant to chafe in order to bring the expected results.

(i) The limits, being put on board and thus clearly visible, are an example of the implementation of the Kanban principle: *"let the rules be simple and visible"*.

6.4. How to begin?

> [...] if you are not reducing your WIP to the optimum level, then you're not doing Kanban.
>
> — Bazil Arden

But how are we to know what good WIP limits for the individual columns are? This is not obvious, especially if you are setting them for the first time. Fortunately, you do not have to guess the correct value immediately.

Since Kanban as a whole relies on improvement through constant change, the WIP issue can be solved this way, too. So, for starters, let's choose arbitrarily high WIP values – for example, let the WIP for each column be equal to the number of team members. Chances are this will change little in the team's work, because such high limits will not force the team to change the way in which it is performing tasks. All right, let it be so at the beginning. During the next retrospective[2] (or earlier, when the team decides it is time for that) lower the WIP limit for selected columns. A good idea is to do this for the columns where:

1. we can see a significant "reserve" (i.e. the typical number of tasks was significantly lower than the WIP for this column),

2. we are especially interested in streamlining work.

---

[2]More on team meetings in chapter 9.

Life will show what happens next. Most likely, the current model of work will collide with limitations such as the WIP limit. At this point, team members will have to discuss how to work on tasks to stay within the limits. Such situations should not be frequent (I am assuming that the limits are still not very painful), and that's good. Otherwise, the team could lose a lot of time debating the new work strategy, or even become frustrated and say: *"I've got nothing to do, because these silly limits do not let me"*. On the other hand, even individual, single cases of reaching the limits will enable team members to work out new approaches and to learn how to divide the tasks better, support each other on tasks, work in pairs etc.

In addition, some metrics can be a signal that the WIP should be lowered[3]. For example, by keeping the statistics on the number of cards in individual columns you can easily find candidates for WIP lowering.

(i) In fact, saying that the right WIP level has been reached is a tough call. It may be *right* now, and then no longer tomorrow, for example because the nature of the tasks the team is working on changes, or perhaps because its environment changes. Do not become too attached to the current decisions in this regard. These are just numbers that are meant to help the team work more effectively on its tasks.

6.5. WIP reduction – how to do that?

Reducing WIP is one of the goals of Kanban. Thus, it is hard to imagine a team that bypasses this element altogether and works on tasks without taking into consideration how many others are in progress. In fact, this should be impossible – if all the tasks have their cards and are represented on the board, an inflated WIP will be clearly visible anyway.

The question is: is it really necessary to set rigid WIP rules? Perhaps it will be enough to scrutinize the board during daily meeting for the number of cards, figuring out whether a reasonable number of cards has not been exceeded? And, if that's the case, deciding which tasks are moved back (e.g. to the *ToDo* column) and which ones the team will focus on (e.g. by working in pairs). My experience shows that it is possible to act either way. As usual, all depends on the context and the team's preferences.

---

[3]More on this in chapter 11.

On the other hand, we all know that some pressure helps us climb heights we would not reach otherwise. They say only challenges make us grow. This is the reason why I would suggest taking advantage of the opportunity for growth that WIP limits are. It's not about tightening the screw or about daily WIP reduction record setting. It is about a certain pressure, about new challenges.

> All the WIP considerations make sense only when the board includes **all the tasks** the team is working on. It could be the case that the board only shows what the team members see as "real work", with other, perhaps numerous tasks not being represented by cards on the board. In such a situation WIP reduction is a sham action, which has no impact on the team's real work.

6.6. WIP for the entire board

So far, we've been assuming that the limits are set for the individual columns. But, if we think for a while what the ideal WIP value is, we might come to the conclusion that the overall WIP for the board should equal exactly the number of people in the team. This is a logical approach – after all, every person can only work on one task at any given moment.

That's logic, but experience teaches us that it is a good idea to leave some room when WIP limits are determined. Whether we like it or not, blocked tasks, or tasks that *"X will test right away, just let him finish something, it won't take him more than two hours"* are on the board, enlarging the WIP. Of course, you can try to deal with each such case. You can also conclude that a certain number of such tasks is not harmful.

The situation is getting complicated when we remember that WIP includes the tasks found in the *Done* column. These are tasks that nobody is really working on – they are simply waiting until the number of cards in this column exceeds a certain critical value, at which point they will all be deployed to production[4]. If there are several WIP "points" in this column, an all-board limit equal to the number of developers will make it difficult to divide the work on the remaining points in a way that will give each team member something to do.

---

[4]The exception here are teams that deploy continuously – in that case the code that passes tests hits production immediately, and the *Done* column is not really needed.

Personally, I believe that column limits work better than one global limit for the entire board.

It is worth noting that team members (typically) do not all have an identical skill set. For example, in a certain team it may be obvious that only four out of seven team members will work the tasks in the *Test* column. In such a case, it seems reasonable to set limits for individual "groups" (separated by skills). Since the tasks pursued in different columns typically require workers with specific skills, setting limits at the column level seems like a reasonable compromise.

Chapter 7. About the cards and the board

The board and the cards are the two key visual elements used in Kanban. It's a good idea to look closer at them.

7.1. Electronic or physical?

Most teams use bug tracker or issue tracker apps to monitor tasks or errors. These are very useful tools, but it seems that, paradoxically, a simple physical board has many advantages over them. I'll try to describe these.

(i) Of course, if you are working in a distributed team, an electronic board is a blessing that you should take advantage of!

First of all, physical objects have a clear, if hard to describe and explain, advantage over virtual ones. Forget the pretty app that allows you to move the cards on a virtual board as if they were real cards. Forget *drag & drop* and the beautiful graphic effects. A screen is a screen, while a physical, tangible and real board is something entirely different. If, until now, you have only been using electronic task management tools, you may find this hard to believe. I'm sure that as soon as you introduce a physical board, the difference will become clear.

- Issue trackers have unlimited capacity – you can easily put several thousand tasks into them. Of course, this makes no sense and means that the team has an urgent problem to solve. ...... and this problem is not the implementation of all these stories! On a board, you can clearly see when there are too many tasks. The team (and thus the project owner) will be **forced** to undertake a reasonable number of tasks.

- The board is great at communicating the team's current work. Remember, human beings remember much more of what they see than of what they hear (and even more of what they both see and hear). Hence a simple conclusion – organizing a team meeting next to the board will significantly improve communication in the team, and the fact that the board is always in sight will cement the knowledge of what the individual team members are doing.

- Using a board, it is difficult to *"sweep tasks under the rug"*. Removing a task from the issue tracker is the matter of one click, and nobody will pay any attention to the emails that inform about this. On the other hand, the mysterious disappearance of a card or its going back to an earlier stage will obviously be noticed. Thus, the board will force the team to discuss problems (*"so what are we doing with this task? is it still relevant?"*) rather than hide them.

- Problems stemming from cooperation with other teams, departments or companies also become clearly visible. Let's imagine a task that has been residing in the very same place on the board for many days. Every day during the meeting, team members say how teams or companies they are working with are the reason why the task cannot be finished. What does the project owner do in such a case? Clearly, he is going to try to fix things to make it possible to conclude the work. And that's the whole point!

- Unlike the screen, the board is an excellent meeting space! The team's life is centered around it. This is where daily meetings are held. If the board you are using is a simple dry erase board, you can expect team members to hold discussions next to it. After all, the other side is unused and it just begs to be used for sketching a diagram while talking to a colleague.

Does this make the issue tracker redundant? No, surely not. Its advantage over the board is that it has room for a precise description of each task. The team should decide about the level of detail and the type of data that will be put there. Let's just say that the task number from the issue tracker should be used in code repository commits. This will help us understand in the future what task the given code changes are connected with.

Experience shows that such "double accounting" is neither a problem nor a burden.

7.2. Cork board or dry-erase board?

If you can, choose a dry-erase board. There are several reasons for this choice:

- dry-erase boards tend to be large,

- the other side of a dry-erase board can be used to sketch diagrams during discussions,

- cork boards are usually affixed to walls; in the case of dry-erase boards you can choose a "mobile" model, which can be moved to the most convenient place at the given moment,

- it is easier to move cards with magnets than cards attached with pins,

- in general, it is easier to make changes – e.g. on a cork board, you must somehow attach a line that delineates a column; on a dry-erase board, you can simply draw it with one motion of your hand.

However, if life forces you to use a cork board, do not worry. You will succeed just right.

Regardless of the type of the board you opt for, buy the biggest one that can reasonably fit in the team's working space.

Where to get the cards?

It's easy. Buy colored paper and cut the sheets in four. Voila! Your cards are ready.

If you get tired with scissor cutting, invest in a guillotine... no, please, no associations – I meant a paper-cutting guillotine!

Large Post-It notes can be found in stores, but I do not recommend them. They have a tendency to bend.

7.3. Final column

> Out of sight, out of mind.
>
> — Common knowledge

The columns on the boards you've seen so far look more or less like this:

TO DO   DOING   TEST   DONE   PRODUCTION

Perhaps, while reading the previous chapters, you were wondering about the rationale for the final column on the board. Wouldn't it be enough to simply remove the cards from the board rather than move them into the *Production* column?

Well, I strongly do not recommend removing cards from the board right after we have concluded the tasks that they represent. There are at least two reasons for that:

• First, other people may also be interested in knowing what tasks have recently been concluded (or deployed to production). They are easy to notice while they remain on the board.

• Second, it is a good idea to give the team a chance to discuss the recently-completed tasks. The retrospective will be a good moment for such a conversation.

7.4. The full picture

It is a good idea to make sure that the board includes **all** the tasks that the team is working on. Sometimes this seems unnecessary, especially in the case of tasks that appeared suddenly (e.g. bugs or tasks like *"the manager is requesting a database report on client X"*) and were concluded quickly. Why make the effort to create and attach a card (right away in the *Done* or *Production* column)?

The reason is that such a card can be an important element of the retrospective. It's great that the team has quickly coped with an unexpected problem in the test environment, but perhaps you should also discuss how to avoid such problems in the future? If there is no such card, it is much less likely that such a discussion will happen. After all, who will remember an event that happened a few days ago, given that so much else has happened since then? It's a pity, because this might be a topic worth discussing.

It happens that a task that seems trivial at first (*"done in half an hour"*) turns out to be somewhat bigger when work on it begins. So it is safer to create cards for all tasks, even if they do not seem to merit it.

(i) All means all. One of the team members is interviewing a candidate? Great, put a card on board. Somebody else is preparing a company-internal training? Kudos to him, and now let's put a card on board, please!

7.5. Shared board

It may be the case that there is a need for one board to include tasks concerning different projects. There may be different reasons for that. Perhaps the project is growing in a way that leads to subprojects being spun off from it, which are still pursued by one team. Perhaps some of the team members are working on an internal project from time to time. Or, perhaps, the office is so crowded that there is not enough room for another board.

Whatever the reason, this happens. Let's think about what to do in such a situation.

This is clearly not a good situation. When the board includes a mixture of information on different projects, it is less clear. It's a good idea to keep that "mixing" to a minimum. You can use different card colors for each project (as long as we have not decided to assign different meanings to different colors). Another idea is to visually separate the board into two parts, e.g. by drawing a horizontal line to separate the cards for different projects.

The optimal situation is, however, for each board to serve one project. You should treat the presence of the other project as a transitional state, striving to put it on its own board.

7.6. Size of the tasks

The size of the tasks represented by the individual cards should be chosen in such a way to make it possible to observe changes in the board as days pass.

If the tasks are too big – for example, if completing them takes two weeks – the board will be still. As a result, the team will be deprived of a positive stimulus that comes from observing the movement of the cards on the board. Team members will be heard saying over and over again that they are working on task X. In turn the project owner, looking at the static board, will start complaining that the work is not progressing at all.

A solution that springs to mind is to divide a large task into several smaller ones[1]. This is usually possible and, depending on the task, may take different forms. One of the options is a division by layer, i.e. creating separate tasks for backend and frontend implementation. Another option is to separate the key functionality from services needed for less likely scenarios. We should remember, however, that in course of development work the developer may discover that he must perform work that belongs to another subtask. What to do then? Perhaps the best option is to mutter *"that's life"*, and not be concerned too much.

Another extreme are tasks that are too small. This is less of a problem. In practice, the only problem with small tasks is that people start complaining after a while because they do not feel like constantly making or moving cards (or they will simply not do that). The solution seems to be to group small tasks and to represent them on the board with a single card.

The size of the tasks can be represented by the usage of different colors. This makes it immediately obvious which tasks can be expected to be completed quickly, and which ones will have to be waited for.

---

[1]We are now entering a territory that is far beyond the scope of this book. Please, reach for other books that describe *user stories* in detail.

Examples

Let's assume that our task is to add one field to a form. In order to consider this task completed, it will be necessary to expand the database model, to make changes to object-representing classes and to add a field in the view layer. In addition, appropriate tests will have to be written (for example: it should be checked whether the validation of the values of the new field generates appropriate messages), and deployment to production will have to be done. Well, this seems like quite a lot for one field in a form... I think, however, that such a task constitutes a whole and does not deserve more than one card on the board.

Let's consider another example. The team decides to work on making the application faster. After a discussion, several cards appear on the board, such as: accelerating database queries, code profiling, introducing cache here and there, testing another load balancer, etc. The division into sub-tasks seems natural here, and experience shows that each of them is likely to take more than a few hours. Hence the multiple cards on the board.

7.7. Let's stay focused on the goal

In one of the projects, we slightly modified the rules concerning the placement of cards on the board. We wanted to see the tasks flowing (and to enjoy the progress of the work), but we also wanted not to lose sight of the overall objective that these tasks served.

For example, if an important thing for the team was to create a database backup mechanism, we would put a blue "DB backup" card in the *Doing* column. Blue was used for large tasks that could take as much as several weeks. Now, on that blue card we would stick other ones, representing backup-related subtasks. A large portion of the blue card remained visible, and the cards affixed to it told us which part of the large task is currently being worked on, etc. "Incremental backup", "Full backup" or "Admin console".

As the work progressed, the cards representing subtasks moved to the *Test*, *Done* and *Production* columns. In the meantime, the "DB backups" card remained in the *Doing* column. When the last subtask was deployed to production, we moved the blue card to the final column as well. The task was finished.

Thanks to such an approach, small subtasks were no longer the goals in themselves. We were now able to keep our true objective in sight. We felt that what we were doing made sense. This turned out to be important at that stage of the team's life.

(i) And that's the whole point! Adjust the board to fit your needs. That's what we did, and it worked. Hope that's the case with you as well!

7.8. Marking persons

Some teams like to include on the cards information about who is currently working on the tasks. This can be done, for example, by affixing to the cards magnets with cartoon characters (*Star Wars*, *South Park* or even *My Little Pony* if it happens to be the team's favorite series).

Such additional information on the boards has its benefits. First, some team members may be delighted to have Lord Vader as their avatar.

Second, you can immediately see which team member does what. Third, you can easily see who has too much on their plate. This third piece of information can be especially useful when looking for ways to streamline the team's work.

Another option is to divide the board into as many rows as many people there are in the team. In such a case, each row contains cards representing the tasks that a given developer is working on.

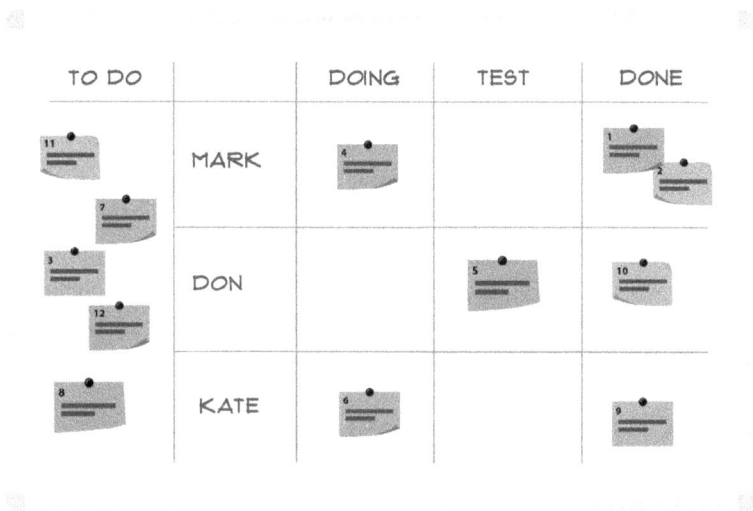

In the case of the *Test* column, when the task is taken over by another person, we have to decide where the card should go. We can agree that it will go to the row of the person responsible for the tests, or that it will remain in the row of the original developer.

On a board like this, how to indicate that Mark and Kate are working on a task together? Isn't a board like this going to discourage cooperation?

7.9. Blocked tasks

Sometimes you just can't go any further. The task has been started, a lot of code has been written, but there is a block ahead. We are waiting for key information from the client. We are waiting for the only JavaScript specialist to come back from vacation. We are waiting for the recently-discovered error X to be fixed, as otherwise the functionality being developed won't work.

A question arises: how should we mark such a situation on board? Most often, two solutions are used here:

- the card is marked with an additional element, e.g. a large black magnet is attached to it.

- a separate space is marked on board, where the blocked tasks are stored.

Personally, I prefer the former solution. Let the card stay where it has been so far – it will be more glaring this way, encouraging team members to intensify their efforts to remove the blocking factor (that is, for example, to check on those responsible for the state of affairs).

In the case of blocked tasks (which are usually known as *blockers*), we can wonder whether they should be counted in the WIP. I believe they should. Primarily because a blocked task is a very bad thing for the team, and it is worth making every effort to get rid of it. The inclusion of such a task in the WIP exerts a stronger pressure on the team, forcing it to work on unblocking it.

7.10. Sudden, urgent and "for yesterday"

There are tasks, and there are **TASKS**. Some of them can quietly wait until it's their turn while others require immediate attention. Others are not so urgent to require immediate work, but they have a non-extendable deadline.

It's a good thing if the cards on the board are marked in a way that makes it possible to immediately recognize their type. There are many ways to do that. For example, for many teams tasks involving errors are a case that requires special attention. A popular way to set them apart is to mark them with red cards. Such cards will clearly stand out on the board, making it clear that the tasks that they represent require a quick reaction.
In turn, for tasks with non-extendable deadlines, the deadline can be written on them.

When discussing the different types of tasks, it is hard not to mention prioritization. It only makes sense to mark types of tasks if they are linked with priorities. We can decide to mark super-urgent tasks with a black magnet on the card, but this must be followed with special (priority) treatment when we are choosing the next task to do.

What will surely make our life easier is introducing as few such markings as possible. Otherwise, we will have to decide whether a task with two magnets is more important than the red error card, and how that compares to a card with a deadline and one magnet.

A visit

Since we have mentioned errors, let me tell a short story about how clear error-marking saved me a lot of trouble one day.

On one project, I was visited by an Important Person. He had an additional task for me. An important one for yesterday. We talked a little, and then I pointed to the board. Embarrassingly, it just happened to be quite red with error cards. One glance at the board was enough for the Important Person to understand that it was not a good moment for adding tasks. His task turned out to be not so important and not so urgent after all. He went away.

Quick path

A popular way to mark tasks that require immediate attention is to create a visually-separated area for them on the top of the board. It's like a special lane for privileged cars. Tasks that are there are to be treated as true priorities, and completing them as soon as possible should be the concern of the entire team.

For example, on the board presented on the following illustration, the error marked as 2 is so important that the team decided to make it the first priority.

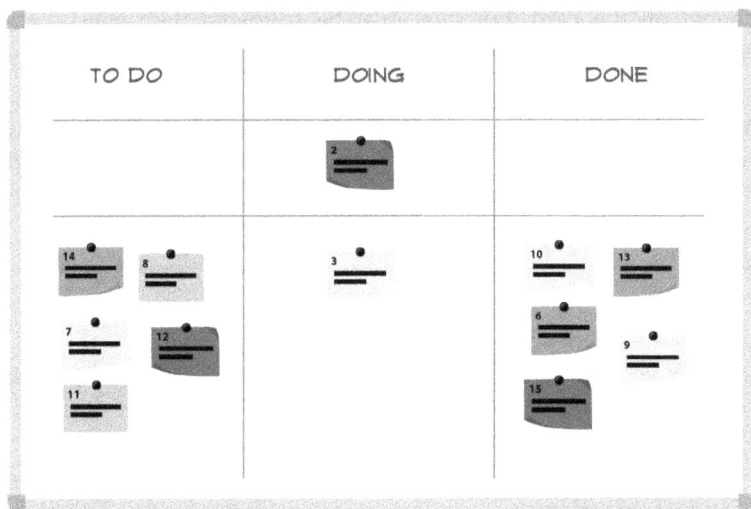

In order to avoid problems stemming from an abundance of tasks "for yesterday", it is necessary to agree with the project owner on how to apply these priority task rules. One example of such arrangements could be a rule saying that at any given point in time, only one task can be in the fast lane.

7.11. Add-ons

The board has quite a large capacity, so there is no reason why the team shouldn't be able to use the empty space as it sees fit. For example, the following information could be put on the board:

- the number of production deployments – e.g. in the form of red dots somewhere in the corner of the board, which would be erased during each retrospective,

- the number of days since last major production incident – with the counter increased (or reset) during daily meetings.

This type of information can also stem from the team's specific customs – e.g. celebrating every 10 days with no production error etc.

(i) I am by no means encouraging you to forcibly stuff the board. Whatever you are adding, let it add value for the team rather than reduce the readability of the board.

On the other hand, there are elements which are unlikely to fit on the board. For example, it makes no sense to put cards representing backlog tasks on it. In most cases, this would be physically impossible.

7.12. Drawings

The cards and the board are a great place for talented drawers. In any case, it doesn't require much skill to add a symbolic "cylinder" to a card with a database-related task. Or to express the emotions related to an error by drawing a skull and bones.

The more talented individuals could, for example, draw a Spider Man hanging down on the line that separates two columns.

As long as the team is interested in making the cards and the boards more lively with drawings, that's great. Let the board come to life, and let its elements stick in memory thanks to funny doodles.

Chapter 8. Let's do it better

> …we are what we repeatedly do. Excellence, then, is not
> an act but a habit.
>
> — Aristotle

Nothing is perfect. Perhaps it once was, but then the conditions changed suddenly (in an unexpected, sneaky way) and perfection stopped being perfect. There are still things to improve, to fix, to refine. The Kaizen improvement philosophy encourages constant efforts directed towards the creation of higher-quality things[1].

Of course, we can promise ourselves that we will write many tests and do code review, that we will create a new environment where the code will mature before deployment, and that we will relieve our QA of all other tasks so that he can focus on the most important tasks, etc. I am afraid, however, that we will end up like the person who decides to quit smoking the next day, to start biking to work, to replace hamburgers with salads and to go to sleep at 10 p.m. Old habits are hard to eradicate, and few of us have enough willpower to introduce revolutionary changes and maintain them. In the case of a team this is even more difficult, since a group always includes people who oppose a revolution more or less actively.

In order to avoid the well-known effect whereby exuberant New Year's resolutions are broken, for Kanban a gradual introduction of small changes is recommended. There are several reasons for that. First, small changes generate smaller resistance. Second, when we make small steps, it is easier to observe which of them take us in the right direction and which lead us astray. This makes it easier to give up bad ideas. Third, the ability to observe how each change impacts teamwork is priceless feedback which can help steer the next improvements in a previously unimaginable direction.

8.1. Mending holes

Many improvements that a team introduces represent immediate necessity. The team reacts to the current situation and introduces improvements. Let's assume that a client has discovered a serious bug in our software.

---

[1]You can read more about the Kaizen philosophy at https://en.wikipedia.org/wiki/Kaizen

A quick discussion leads to a decision on what to do prevent similar errors from hitting production. One of the tasks on the continuous integration server ends in an error for the second time this week due to lack of resources (e.g. memory). A decision is made to use a stronger machine (earlier configuration trials turned out to be insufficient). And so on.

Such decisions can be made practically at any point. They do not require any "official" team meetings; frequently, a handful of team members will be enough to make important decisions.

8.2. Optimization

> It is said that improvement is eternal and infinite. It should be the duty of those working with kanban to keep improving it with creativity and resourcefulness without allowing it to become fixed at any stage.
> — Taiichi Ohno *Toyota Production System*

Reacting to current events is one thing, while regular striving towards perfection is something altogether different. The retrospective, which we will discuss soon, seems to be a good moment to discuss improvements that the team could introduce. They can be very different; the important thing is for them to be aimed at quality improvement. This might entail a decision to use the newest version of library X, but also a decision to stop writing Javadocs (hurrah!). The important thing is to give all team members an opportunity to submit their ideas[2] and to ensure that they are all discussed in a respectful, understanding atmosphere. It is also a good idea for the team to agree to a limited number of improvements on one occasion (e.g. just one), which greatly increases the likelihood of them being actually implemented.

One form of a planned introduction of improvements are the so-called "Kanban kata". Here's the idea. The team chooses an objective. It can be a very ambitious goal, perhaps one that seems unattainable at the moment. What matters is for this goal to be relevant for the team. Some examples include "zero errors in production", "great, constructive meetings", "cutting lead time[3] by half" or "service availability at 99.999%".

---

[2]In addition to a discussion, which often ends up dominated by the most outspoken individuals, there are several techniques that enable all team members to express their opinions. This, however, goes beyond the scope of this book.

[3]We will discuss *lead time* in chapter 11.

After choosing the goal, the team chooses intermediate steps whose realization will bring the main goal closer. These intermediate goals must be well-defined, measurable and feasible. In order to attain them, the team must introduce certain changes to its work organization and/or technology. In line with the previous discussion, in order to make these changes more likely to succeed, they should be evolutionary. Implementing them at certain time intervals will give the team a chance to draw conclusions and to consider further changes.

For example, let's consider the previously-mentioned goal of "zero production errors". Perhaps a goal that ambitious should be thought of as a remote mountain, whose outline is barely visible on the horizon. Seeing how far it is, the travelers may wonder if they are ever going to reach it. But as they move towards it, they can be sure they are going in the right direction.

Similarly, a team of programmers that is moving towards "zero errors" will make many significant improvements. Perhaps it will start with an obligatory code review for all the tasks. The next step may be the inclusion into the building process tools that generate code quality metrics. At yet further stage, an environment may be developed in which strong end-to-end tests may be run. Another improvement may be the introduction of pair programming (e.g. for key functionalities). And so on. It is doubtful that all these actions will eliminate all the errors, but I am convinced that they will eliminate many of them, leading to improvements in quality and productivity.

There may be some difficulties with implementing the "Kanban kata" idea. There are no precise rules, and there are many aspects to consider. For example, it is unclear what criterion to use when selecting a goal to pursue. And should it be one goal, or several goals pursued in parallel? Measuring progress in attaining the goal can be another difficulty. Sometimes, metrics will help here (we will discuss them soon in chapter 11). But for some goals, it will be difficult to come up with any measurable values[4]. For example, this would be the case with the previously-mentioned "constructive meetings". The team must also decide how it is going to introduce improvements.
One thing seems sure, though. Entering the "Kanban kata" road will require many discussions and decisions.

[4]Douglas W. Hubbard's book titled "How to measure anything" offers valuable insights on this problem.

Discuss

In the process of creating software, the room for improvement is virtually infinite. New tools and libraries appear, or perhaps new versions of the existing ones. The programmers' fingers are always itching to refactor and refactor (which is always accompanied by the authentic conviction that this will soon bring perfect code that will never require any further changes). The manager is moving boxes on the Gantt chart in the (vain) hope of accelerating work. Testers, QAs and DevOps are automatizing as much as they can so that we no longer have to do anything manually. In short – we are optimizing.

Having said that, it would be good to limit our drive for perfection somewhat. Not every improvement makes sense. Sometimes, it is better to leave the things as they are rather than waste resources on costly improvements that bring meagre profits. A good way to find truly valuable improvement ideas is to discuss them during a team meeting. There will certainly be someone who will question the ideas of other team members, provoking a (hopefully constructive) discussion. And that's the whole point.

Chapter 9. Team meetings

> Blessed by the one who, having nothing to say, does not put that fact into words.
>
> — Julian Tuwim

Team members often sit at adjacent desks, have their meals together and lead rich company-internal social life that includes the sending of funny cat pictures on the chat. Experience shows that despite all these activities, good teamwork requires meetings that focus on the project being pursued. The best meetings are constructive ones, which have a clear goal and a reasonable time limit.

Kanban does not mandate any meetings. The practice shows, however, that it is a good idea to hold them.

In the teams I worked on, we held regular meetings. Every day we spent a moment discussing current topics, and upon conclusion of a certain period of work we would meet to discuss our achievements and the possible improvements, after which we would start planning the next period of work.

(i) If you are wondering what meetings you need, then, in addition to reading this chapter, I recommend getting familiar with *"how it's done in Scrum"*. A good introduction would be reading the Scrum Guide.[1]

9.1. Daily meetings

The **daily meeting** is a meeting which probably everybody knows, and which serves for team members to briefly describe:

- what they've done,

- what problems they have faced,

- what they will be doing now.

---

[1]The Scrum Guide, or the official Scrum manual, can be found at https://www.scrumguides.org/

Thanks to this information, team members update each other's knowledge of the progress of the work, possibly catching problems that may be blocking further progress.

In the case of Kanban, daily meetings happen at the board. As a result, all team members can see, and not just hear, what the progress is on the individual tasks.

Except for this, these meetings hardly differ from those proposed by the Scrum methodology. They are usually standing meetings and take a few minutes, with any discussions resulting from the problems signaled by team members happening later, without everybody necessarily present.

> (i) It is imperative for team members to update the board before or during the meeting (by moving the cards to appropriate locations).

9.1.1. Focus on tasks

There is one important change that is worth introducing into the daily meeting scheme. In classical Scrum, each team member discusses their achievements, problems and plans in turn. This gives a picture of the work of all the individuals, without necessarily telling us anything about the totality of the achievements of the entire team – and this is the piece of information that seems more important from the point of view of the client (or project owner).

In order to get information about the progress of work, another approach should be tried. Developers should describe the progress of the work on tasks **card by card** starting with the right-hand side of the board.

A story might look like that:

- *Production* column: **Dev 1** (to the project owner): 123 went to production, clients should be notified that they finally can start setting limits for individual entries. In addition, we have fixed cache configuration, so the administrator's panel will work faster.

  or

- *Done* column: **Dev 1** (pointing to a card marked 123 on the board): We haven't deployed the newest version to production, because we are waiting for 123 to be ready.

Dec 2 and **Dev 3**: Stories tested: adding and removing comments. We are waiting for notifications about new comments under the text to deploy everything.

- *Test* column: **Dev 4**: Comment notification is being tested. It looks good, I will double-check and let you know when it's ready for deployment.

- *Doing* column: **Dev 1**: I am working on automatic category suggestions. It's an uphill battle, I am buried in refactoring in the notification module (the entire team sighs understandingly).
Dev 5: I am working on changes to CSSs, I should pass this for tests today. When I am done, I'll start on fixes to the drop-down menu (from *ToDo*).

- *ToDo* column: **Project owner**: I've added a card for the logging-out problem that was reported yesterday. It would be good to close it by the end of the week, but no need to panic.

As you can see, each task is discussed by the person who knows the situation the best. Frequently, several team members talk about the situation in each of the columns. If some team members have been working on more than one task, they will speak several times during the meetings.

Experience shows that such a story is likely to give a better picture of the wok on the project. It also has several advantages:

- Thanks to the focus on tasks it avoids the constant repetition of the same things when several people have been working on one task.

- It is hard to construct stories about God-knows what when there is a single card to discuss. In the case we are discussing all the achievements of the preceding day, some of us tend to speak too long about nothing[2].

- All the tasks get discussed. In the case of a large team and multiple cards, it is really easy to overlook the fact that one of the tasks failed to be mentioned by anybody.

---

[2]Most likely due to the following reasoning: *"the longer I am speaking, the more I have done, the more valuable team member I am"*.

At the end of the meeting, it is worth asking whether some team member might have some additional information. This will make it possible to discuss topics that are not represented by a card on the board.

9.2. Retrospective

> When you master the art of the retrospective, you are honing in on kaizen.
> — Jim Benson *Personal Kanban: Mapping Work*

The other type of a meeting, which, lacking a better term, I will describe as a "retrospective"[3] happens less often. Its goal is to wrap up the recently-concluded period of work and to agree on the improvements that should be introduced.

Such a meeting should happen every... Well, it is difficult to say exactly how long the breaks between retrospectives should be. My experience shows that reasonable boundaries are one to three weeks. I suggest to choose some period for the beginning (e.g. two weeks) in order to check if such meeting frequency works.

(i) Some teams do not decide on any specific frequency of retrospectives, but call them on an ad-hoc basis, usually when the team members feel that there are enough topics to wrap up. I do not like this approach. I believe it is a good idea to have a specific rhythm which opens and closes certain stages of the team's work.

There are many great practices, tasks, exercises and games that can (and should!) be used during retrospectives. However, since this is a book on Kanban and not on running retrospectives, I will try to focus on Kanban-specific elements.

9.2.1. About the process

First, especially when you are starting to work with the board, it is worth asking team members about their impressions on the use of Kanban. This can lead to a discussion, especially if:

---

[3]"Retrospective" or a "summary meeting"... everybody is gonna call it "retro" anyway!

- the columns planned do not correspond to the real production process,

- in the period under discussion, there have been cases that did not respect the natural flow of the cards on the board (e.g. tasks that were stopped by a decision of the management),

- the agreed WIP limits have been exceeded.

Such situations may make it necessary to:

- adjust the board to the reality (e.g. by adding a new column),

- change the reality by taking steps that will prevent these abnormal situations from occurring,

- discuss ways in which the team should react to anomalies.

9.2.2. What have we done?

The central element of the meeting is the board, and the discussion during the retrospective will focus on it. We can start from the right-hand side, summing up the tasks that the team has completed since the last meeting.

It is a good idea to celebrate this part of the retrospective. Let everybody have a look at the number of task that have been completed. If there are many cards to be taken off the board, it is a good idea to group them on the board (in the last column, which comprises the completed tasks). By way of example, we can separate groups such as fixed bugs", "new functionalities" and "internal improvements". You can also group the tasks by the project area they fall in (creating groups such as administrator panel" or "search engine").

It is also a good idea to remember the history of some of the tasks, especially if it includes some lesson that should be remembered. It is good not to take the cards off in a rush, saying a few words on each of the tasks. This can lead to valuable conversations.

The "ceremony" of taking the cards off makes the most sense if we have ensured to reflect **all** the tasks worked on by the team on the board.

Taking the cards down is a very important moment of the retrospective. If the just-completed period of work was a successful one, it will be hard not to feel proud seeing the significant number of completed tasks. On the other hand, if the work was not going so well, it will be hard to avoid a discussion of why this was the case and how to improve work efficiency.

9.3. Reaching the heart of the problem

A retrospective is a good moment to discuss the problems that have appeared in the recent period. During the discussion, it is worth stopping over the problematic situations, both the ones that have been solved as well as the ones that are still awaiting a happy ending. In the first case, it may turn out that new ideas have been born in the meantime.

With respect to the still-nagging problems: you can try solving them during a retro meeting, or you can organize a separate meeting devoted to them (perhaps in a smaller group). The key thing is finding the root cause in an atmosphere of searching for solutions rather than for the guilty persons.

(i) Searching for the guilty causes trauma (to the individuals and to the entire team). The search for causes and solutions leads for team consolidation and improvement in productivity.

9.3.1. Improvements

In line with the Kanban-promoted Kaizen philosophy, we should constantly implement small improvements in our work. The retrospective is a good moment to brainstorm and to select the actions that could make our work more effective and satisfactory.

There are many techniques that help the team determine the list of changes to be implemented. Discussing them is outside of the scope of this book, and I will limit myself to a handful of key suggestions[4]:

- it is a good idea to collect the opinions of all the team members,

- work in sub-groups brings good effects,

---

[4]In the C appendix, I've collected information about where to find a detailed description of these issues.

- it is a good idea to replace team-wide discussion with other ways in which team members can express their opinions,

- in order for the changes to be truly implemented, they should be proposed and approved by the team, and not imposed from above,

- a person should be assigned to each planned change that will monitor its implementation,

- during the next retrospective, it is worth discussing the effects of the actions undertaken in the preceding period,

- out with monotony! - it is a good idea to find different types of exercises,

9.3.2. Beware of monotony

> My life is very monotonous [...] And, in consequence, I am a little bored.
>
> — Fox from the Little Prince

A well-run retrospective can do lot of good to the team's work. There is another reason for which it is worth holding. Let's think about what the rhythm of work in a Kanban-based project looks like. It is simple. New tasks appear in the *ToDo* column, the team takes them up and realizes them. Then new tasks appear and the team works on them. Then more tasks, and more tasks… In a word: monotony.

The Sprints we know from the Scrum methodology have their disadvantages, but they certainly have the clear advantage that they set an unmistakable rhythm. The end of a Sprint is always a "finishing line" of sorts that everybody is striving for together. It is also a moment to celebrate success or to discuss failures. In the monotonous flow of tasks that is typical for Kanban, there are no such clear stops. It is thus worth making sure that such stops appear in the team's calendar.

What is more: it would be good for meetings to differ from one another. It is a good idea to vary the retrospectives by introducing various activities so that they always bring a surprise of sorts.

ⓘ The team's efficient work largely depends on how its members are feeling. Monotony dampens enthusiasm and morale. Do not let it sneak in.

Combining meetings

There is a temptation to combine meetings, simply just to save some time. After all, planning can be combined with a retrospective. Likewise, a daily meeting can be basically skipped on the day of the retrospective, since in a few hours we will be discussing all these cards during a retrospective. Yes, you can do this, but I do not recommend this.

There was a moment when we tried to minimize the number of the meetings this way. After a while, however, we separated them and since then we have been holding them separately. This experience taught me that there are important reasons to keep the meetings separate. In my opinion it is not a good idea to combine daily meetings with the retrospective or the planning meeting, because:

- a constant rhythm should be maintained, with the team discussing the progress of the work at the same time every day,

- when discussing many important issues during the retrospective or the planning meeting, there may not be enough time to focus on the current progress and problems of the team.

I also do not recommend combining the retrospective and planning for the following reasons:

- the meeting grows too long and thus becomes tedious and ineffective,

- some participants (e.g. the product owner) may be interested primarily in planning, and others primarily in the retrospective,

- the accumulation of topics during one meeting may lead to the individual parts being "watered down".

This is why I would recommend organizing separate meetings of all types. It will not be a loss of time as long as each meeting is well-prepared and well-run.

9.4. Planning

Kanban does not impose any specific way to plan tasks. From the point of view of the team's current needs, it is imperative for the *ToDo* column to always contain enough tasks for the persons who complete work to never be left empty-handed.[5]

Another important thing is for team members to know the size of the tasks planned for the upcoming months. First, this will let them better organize the current work, i.e. in a way that prepares the ground for upcoming changes. Second, working with the awareness of the goal is more engaging than working on tasks whose deeper sense remains unknown.

The teams I've worked on usually held planning meetings on the same days as retrospectives. That's not surprising given that the conclusion of a certain period is a good moment to wrap it up (retrospective) and to look ahead (planning).

I will not discuss planning methods in depth here, because this book is not about them. Just as in the case of retrospectives, I will merely attempt to shed some Kanban light on them.

In the case of Kanban, planning can be started with a look on the tasks that can be found on board, especially those not yet completed. It is a good idea to discuss all of them, thinking about how to deploy them to production as soon as possible. From time to time it may turn out that the given task is no longer that important for the project owner, and some part of it is not even worth working on. A board review ends on the column that contains the tasks that have not been started yet. This is where the planning phase really starts. How it goes depends very much on the specifics of the team's work, so I will not offer any suggestions here. I will merely say that from the point of view of Kanban, it is important to:

- consider whether the tasks currently found in this column still deserve to be here,

- not allow for this column to be overwhelmed with too many tasks; the column's physical size may indicate a reasonable number of tasks.

---

[5]Typically, the converse is the case, i.e. there are too many tasks to do. There is always a lot of work.

(i) It is a good idea to encourage all team members (all of them!) to think **ahead of** the meeting what tasks they believe should be worked on in the near future. This will make it possible to discover different tasks, not just the ones that the project owner is keen on.

9.4.1. Priorities

Of course, it is a good idea for the developers to know which of the tasks amassed in the *ToDo* column should be worked on first. Thus, it is advisable to encourage the project owner to express his or her priorities. In addition, priorities should be clearly visible – this can be easily expressed on the board by arranging the cards in the *ToDo* column appropriately (e.g. by putting the most important tasks on top).

We can also introduce some easy rules that determine the order in which the tasks are done. For example, there could be simple rule: *"bugs are fixed first"*.

It is certainly a good idea to prioritize if the project owner is not constantly available. Otherwise, it is enough to ask them which of the tasks is the most important one at the given moment.

My experience shows, however, that setting priorities often has little impact on the order in which the tasks are worked in. And that's not because developers prefer to pick the best bits and to leave tedious tasks (though this is certainly the case). The real reason is that some tasks match some people better than others. Even if the team comprises people with similar skills, they are never identical, fully exchangeable elements of the machine. One person deals better with multi-thread code, while another one wrote the given part of the system and knows it inside out. Thus, sometimes already at the planning stage it is obvious that task X will be taken up by Peter, and task Y by Jerry. And thus, when Peter has to choose a task from the *ToDo* column, he will most likely choose task X, even if task Y is higher on the list of priorities.

(i) Exchangeability among team members is a good thing, and certainly it is a good idea to invest in techniques for propagating system knowledge that support it (pair programming, peer code review etc.). I maintain, however, that it is difficult to achieve in practice.

9.4.2. Estimating task completion time

Project owner always wants to know when a given task will be completed. That's understandable. He is not being mean, he really needs that information. Hence a tendency to estimate the time a task will take, in order to be able to announce: *"this will be ready in 3 days"*. A planning session seems like a good moment to meet this wish. Isn't knowledge about when a given task can be expected to be completed a good input into the planning process?

The history of a certain estimate

In one of the projects, we attempted to estimate the number of hours that each task will take. We would put this information on the card – in the upper right-hand corner. After completing the time, we would put the true time spent on it. In order to facilitate the determination of the actual time needed to perform a task, we would write down "intermediate times" at the bottom – i.e. the number of hours spent on the task on given days. The cards looked as follows:

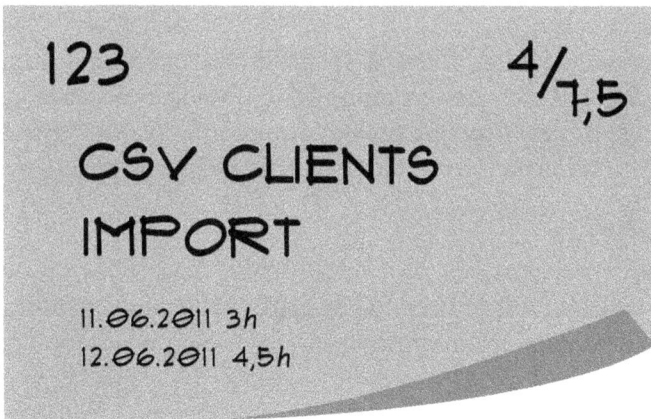

After many weeks of estimating, it turned out that our estimates were still not worth much. The spreadsheet where we meticulously entered the estimated and the actual times showed clearly that in spite of all the efforts, we were wide off the mark. Well, like many other teams before us we discovered that estimating hours is difficult!

Estimating hours is difficult indeed, but that's not the point. The key question is whether such estimates are truly needed. If the team does not promise to execute a certain portion of tasks in a specific time (as is the case in Scrum Sprints), estimating the time to complete a task becomes practically unnecessary. For this reason, many teams avoid giving any numbers that say *how much this is going to take*. In chapter 11 we will discuss metrics and we will find out that even without precise estimates, it is possible to say with a good deal of certainty how much a given task will take based solely on the effects of the team's earlier efforts.

(i) Estimating is costly. Usually it is done by the entire team during a session that takes... one hour? two hours? three hours? Or perhaps the entire day? If you multiply this by the number of hours, the resulting number of person-days can be scary...

Loosening working time estimation requirements for a given task, many teams opt for simple measures which only describe the relative size of the tasks. This means that a label attached to a task is not meant to tell us when it will be completed, but it allows us to put it in a certain group of tasks of similar "gravity".

Clothing-related descriptions are popular, ranging from S through M, L to XL. We can also use a simpler scale, dividing the tasks into small, medium and large. Another option is estimating on the Fibonacci scale, where increasing numbers indicate the increasing uncertainty with respect to task size: 1, 2, 3, 5, 8, 13, etc.

It is important not to be fixated on determining task completion time (which will not work anyway, but is time-consuming and frustrating), but to try to determine relative task sizes.

(i) Experience shows that some of the tasks whose completion time had been estimated ended up never being implemented. In the Kanban understanding of the world, this is an unnecessary energy outlay which should be avoided. Which is what I was suggesting above.

> ### But discussions are priceless!
>
> I was critical about time estimation sessions, but I must admit that they have a certain very valuable consequence. When we try to decide whether a task is going to take 4, 5 or 8 days, it may be necessary to discuss the topic in depth. This rarely leads to precise estimates, but it does allow the entire team to get to know the task. Frequently, interesting questions are discussed on this occasion, and certain doubts are clarified which, if left untouched, could prove to be very costly afterwards.
>
> When opting for less detailed estimates (e.g. by choosing the "shirt" sizes S, M, L, XL) we should remember not to omit a group discussion on details. Perhaps it should come before a task is moved to the *Doing* column.

9.5. Let's not waste time

Let's start with a reflection-inducing joke: *"A programmer is someone who deletes emails and avoids meetings"*.
Very honest and, unfortunately, true. Well, let's be honest, too:

An ill-planned and ill-ran meeting is a terribly frustrating waste of time!

The issue of effective and engaging meetings is far beyond the scope of this book. But it is so important I had to mention it. I recommend an investment in the meeting facilitation skill to enjoy effective meetings.

Chapter 10. The board isn't everything

> True excellence is a product of synergy.
>
> — Mack Wilberg

The board may be a nice thing to have, but let's not forget that it is just one of the many elements that support the team in its journey to perfection. It will show its full splendor only when it is accompanied by other elements and practices, such as constructive meetings, good communication inside the team or close cooperation with the project owner. Efforts aimed at code quality maintenance, from code review to continuous deployment, are no less important.

Below, I will discuss two additional elements that support the board.

10.1. Information radiators

Information radiators are all the different methods of conveying information in a way that makes it literally impossible not to notice.
Imagine a large screen used to project statuses from a continuous integration server – green for correct, red for errors. Can you walk past it without noticing that three of the forty recently built projects ended in a catastrophe? Imagine a loud gong that rings after each successful deployment in production. Nobody who is around can fail to hear it. Imagine a red emergency light (like the one you see on an ambulance) which, accompanied by a loud sound, lights up for a few seconds when the production environment monitoring system raises an alarm. It is simply impossible to escape such an intrusive method of delivering information.

In a way, the Kanban board can be seen as one of these *information radiators*. It's large and easy to notice, and it makes all the alarming situations in the life of the team more visible. You immediately notice if there are too many cards in one column, or if there are very many red cards (which denote bug-fixing tasks). In this way, the board constantly reminds you of certain problems, and it will continue to do so until the problems have been tackled and dealt with.

(i) There's a difference between knowing something and being reminded daily and invasively of the existence of this "something".

The board shows very clearly what the team is busy with at a given moment, and how advanced the individual tasks are. A great way to complement this information will be the screen that displays information from the continuous integration server. In this way, the data on the current and future tasks will be supplemented with current project build status.

(i) I guarantee that the team will put a lot of effort into ensuring that the daily meetings are accompanied only by the green color from the build server.

10.2. Definition of Done (DoD)

Most likely, your board includes a *Done* column, but what does *Done* actually mean? When is a task done? When it has been implemented? Committed to the code repository? When all the unit tests are passing? Or do we also require integration tests, end-to-end tests and so on? What about code review? Is it necessary to conclude that the work is done? What about documentation (if you can be bothered to write it at all)?

You might believe that it is a non-issue because it is obvious when something is *Done*. You may find out, however, that individual team members have different opinions on that. This can happen especially when the team is fresh, or if it includes some members with limited professional experience.

In the case of some of the teams that I worked with, determining the *Definition of Done* (DoD) was of great importance, allowing us to improve the final effect of our work (i.e. the tasks checked as *Done*). In the case of other teams, which were well integrated and comprised experienced experts, there was no such need at all. The question is, which group would your team fall into? Whatever the end result, the criteria for marking a task as *Done* are worth talking through.

What to do when DoD has been established? It is certainly worth recording, for example on the project's wiki. My recommendation is, however, to make a large poster with a very concise and very clear description of the DoD elements you have agreed on. The poster should then be hung in a highly visible spot, so that it sinks in.

In the case of the Kanban board, we must remember that what should really matter to us are the transitions between all the columns. The *DoD* definition that is taken from Scrum must therefore be expanded, so that each team member is aware of the criteria for completing a task in the context of each individual column. What must be done for a task to be transferred from the *Analysis* column to the *Doing* column? When can a card leave the *Doing* column? What must be done to move it from *Test* to *Done*? The final *DoD* is a sum of all these minor rules.

Chapter 11. Metrics

> The more metrics there are, the better. One is always going to increase and you will have something to boast about to your supervisor.
>
> — The Great Book of Cynic Managers

Kanban does not force us to make any measurements. However, in order to evaluate the effects of the changes in the team's way of work, it is good to have numerical data on hand. Metrics can also prove helpful in estimating tasks.

We will start the discussion of metrics with the most popular ones that are commonly accepted in the Kanban world. It is worth noting that *"commonly accepted"* does not mean that the metrics have a precise definition. Just the opposite! On the web, you can find numerous blogs that define them in many ways. The variant presented here seems to be popular (and reasonable) enough that I've decided to avoid the discussion of other options.

11.1. Lead time

One of the most commonly used metrics is the so-called *lead time*, i.e. the time it takes for a task to move from the initial column to the end of the board. For obvious reasons, we are very, very interested in keeping it as short as possible.

Measuring lead time – e.g. by noting down the date when the card was created and comparing it with the date of completion of the work on the task – will allow us to estimate the time when further tasks can be expected to be completed. Of course, if the tasks on our board are of different sizes, we can separately write down the lead time for each type of task. This will allow us to be certain in the future when we declare that *"a medium-sized task, M, will take us 3-5 business days"*.

In some cases it may be interesting to have a look at the actual *development time*. We usually count it from the moment when a task leaves the *ToDo* column. This makes sense, especially if the *ToDo* column is not very stable. This happens when there are frequent priority changes and tasks return from the *ToDo* column to the backlog and are replaced by new tasks.

Another reason to measure development time may be the wish to optimize this element of the software production process.

We can also run a separate *lead time* statistic for errors. The data collected in it may be very useful. It is one thing to give the client hollow promises that a bug will be fixed tomorrow, and quite another to have a number of datapoints saying how much it really takes to fix an error from the moment it is reported until it hits production.

11.2. Delivery rate

Since we have discussed lead time, we can now introduce the so-called Little's Law. It is a law that expresses the relationship between WIP, lead time and delivery rate. Delivery rate tells us how many tasks will leave the process in a given time unit – e.g. that our team of developers completes two tasks per day on average. Little's Law is expressed by the following formula:

```
delivery rate = WIP / lead time
```

By way of example, if our team can work effectively on 6 tasks (WIP) and completing a task takes us 3 days on average (lead time), we will deliver 2 tasks per day on average (delivery rate).

In order to deliver more tasks every day, we can try two things. First, we can shorten lead time – and this is what we should focus on.
There are many options here. Perhaps we can eliminate unnecessary elements of the process, make work on tasks parallel, focus on quality to avoid bugs, optimize the process of knowledge transfer between team members to avoid misunderstandings etc.

Second, we can try to expand the team to allow it to work on more tasks in parallel (increase the WIP). We are aware, however, that adding developers to a team is unlikely to bring desired results[1], and the more open tasks there are, the more likely disorder is to rule. This is not a road to take.

---

[1] As the saying goes, *"nine women cannot have a baby in one month"*.

11.3. Other metrics

The world of metrics does not end on those described above. The question is, *"what metrics are relevant for my team"*?

From my experience I can say that interesting metrics include:

- the number of open bugs, which shows quite well when quality problems occur,

- WIP – the number of open tasks, i.e. the ones that have left the *ToDo* columns, but have not reached production yet,

- the number of completed tasks of each type (S, M, X, bugs), calculated when cards are taken down during a retrospective; it gives us a reasonable awareness of how many tasks of different types can be expected in each iteration.

It is worth stressing here that lead time (as well as other data indicating how many tasks can be completed during a specific period of time) is a good alternative to estimations of the time to perform a task that we know from Scrum. The difference is that estimates are a sort of a guessing game. In contrast, metrics are about facts. As long as the tasks in the upcoming weeks resemble those from the preceding weeks (in terms of size), the data collected will be sufficient to get an idea how many can be completed. We will thus avoid long (and often frustrating) time estimation sessions.

11.4. Attention!

> What gets measured gets manipulated.
>
> — Craig Larman

One thing that is typical for metrics is that people who are aware of them make (un)conscious efforts to boost them. This means that while a given aspect of the software production process is likely to improve, problems may appear in other areas (especially those that are not covered by any metric). Thus, it is advisable to use a set of metrics that will focus on all the aspects that are of importance to us.

It may even happen that a team that is fixated on improving a given metric will start to engage in (self-)deception! By way of example, let's assume that one of the metrics is the *"number of open bugs"*. It is a very commendable metric which is aimed at encouraging quality improvements. In practice, however, it can lead to bad things, such as concealing errors.

> History has seen metrics dependence. Those who suffer from it make meticulous notes of all the metrics, even though they are not using them in reality. A good medicine is teamwide discussion of the usefulness of the data collected.

11.5. Charts

Oh, those pictures worth a thousand words each! Truly, in the era of spreadsheets it would be a seen not to visualize the data we have collected.

It's not much work, and the gain can be big. Let's remember that one of the principles of Kanban is visualization. Making a WIP chart, or one on the number of open bugs over the past few weeks, is no problem. Now all we need is to project it on a large screen that is highly visible to all team members.

Here is an example of a chart that shows WIP, i.e. the number of open tasks, and bugs, i.e. the number of open bugs:

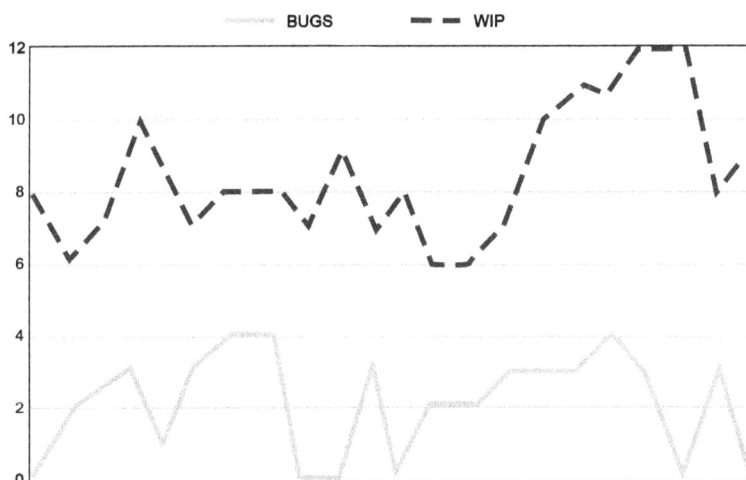

A very popular chart used by teams that work with Kanban is the so-called CFD (Cumulative Flow Diagram). Creating it is very easy – each day, we count the cards in the columns and present them on the board. Completed tasks are presented at the bottom of the chart[2], and the tasks from the *ToDo* column are shown at the top of the chart. Between them are the tasks we count in WIP – i.e. all the tasks that we have started working on but are not yet completed.

Such a chart has one interesting property. We can use it to read the time that the team needed to perform all the open tasks. As we can see, this time can fluctuate significantly as the work on the project progresses.

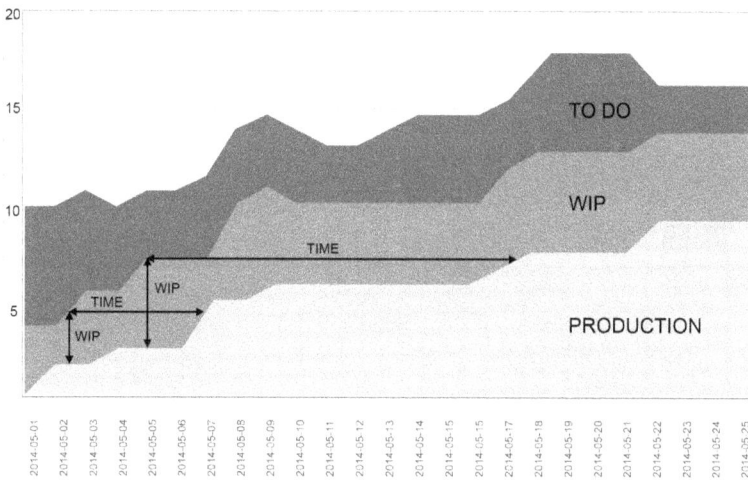

---

[2]This is why this chart is described as a *Burn-up chart*, in contrast with the *Burn-down chart* we know from Scrum.

Variable lead time

There are many reasons why changes to *lead time* are natural. Vacation, sickness and holidays all mean that the number of tasks the team can do in specific periods of time varies. The tasks that belong to one "weight category" (e.g. S or XL) can differ quite widely. Hence, the observed differences in *lead time* should not surprise us too much.

It is advisable to observe the values of this metric over a longer period of time, and not to panic when comparing its values in different time periods (nor to become euphoric). Only an in-depth analysis can show whether the observed fluctuations have natural causes, or whether they indicate (un)favorable changes in the team's work. The key issue is observing a trend and connecting lead time values with the broader context of what is happening in the team.

Chapter 12. Warning signs

Just one glance at the board can tell us a lot about inefficiencies in the way the team is working. Below, I will describe several patterns of cards on the board which should make team members alarmed.

Too many cards

We sometimes deceive ourselves. We convince ourselves that we are able to do more than truly can do. On the Kanban board, this situation is reflected by too many cards. And how many are *too many*? I'll give you an example – if the team comprises 5 members, and the *Doing* and *Test* columns have 15 cards altogether, there are certainly too many cards. It is simply not possible for such a small team to work on so many tasks simultaneously.

This state of affairs can be a consequence of many blocked tasks. They remain on the board, because work on them has started, but no-one is really working on them.

In addition, frequent changes to priorities can result in a significant number of tasks being ditched for some time, while new cards appear next to them on the board.

In theory, a large number of cards should be glaring, but in practice the excessive cards can be hidden – for example, seemingly by accident, one card is glued to another one.

Motionlessness

It is surprising how easy it is to notice cards on the Kanban board that keep stuck in one place. Days pass, other cards fall off like leaves in the fall, and these remain stuck without budging. You may be surprised to discover that such motionlessness is totally invisible in the issue tracker. On the board, you can see it right away. Is this magic?

The sources of the lack of motion can be varied, of course. It may be a consequence of a failure to fully define a task, of its being externally blocked, or of the key developer being on vacation. Whatever the cause, the team will have to face them.

Another cause, and a quite worrying one, may be the team's negative approach to the board, as a result of which team members do not make the effort to create or move the cards...

Returning tasks

A frequent problem are tasks which are found in the *Test* column during one meeting and then back in *Done* at the next one (or perhaps remain stuck in *Test* depending on whether the team decides that the tasks in which bugs are found should return to *Done* or not). If there are many tasks like that, it is clear that developers readily move sloppily done tasks to the *Test* column.

What about the Definition of Done? Is there one at all? And if so, does anybody pay any attention to it?

Ugly doodles

The board is ugly. The cards are sloppily made, unevenly hung, with some of them already fallen onto the floor. Some other ones are hard to read.

The board is not a museum exhibit that must be carefully cleaned every day. No, it's not a problem that one card is twisted. It is not a problem that one of the lines is a little smudged. It is enough just to go to the board, fix it, and all is well again. But the situation is clearly not good when the board is chronically neglected. When it is clear that nobody really cares.

Glaringly red

Is the board simply full of red bug cards? Well, no comments needed, really. Quite clearly the quality of the code that hits the *Test* column (or is deployed to production) leaves much to be desired.

Accumulation of cards

It is also clearly visible on the board which software production stage is a bottleneck. It is a good idea to monitor the number of cards in the individual columns. If there is a regularly occurring accumulation of cards in one of them, something is clearly not right.

A classic example of such a situation is a large heap of tasks hanging in the *Test* column. The cause is usually the fact that developers are not doing enough checks, pushing all the responsibility for quality testing onto the (usually not very numerous) testers and QAs.

It can also happen that tasks are accumulating in the *Done* column. The culprit here is usually the team's reluctance to deploy to production. It is surely important to find out what its source is.

...So, after some consideration, discovering an accumulation of cards on the board should be welcome news. We can finally see where the task flow is obstructed. Let's not deceive ourselves, this has been like this for a while, but now, thanks to the board, it is clearly visible. And since we know where the problem is, we have a chance to fix it.

Chapter 13. Good advice

Ask not, what Kanban can do for you,
ask what you can do for Kanban.

— JFK (supposedly)

You have certainly found a lot of advice on the pages of this book. Some of it was quite explicit, while some has been maliciously hidden between the lines. Below is a list of these pieces of advice that I consider worthy of stressing.

- Do not let the board lie. It should reflect the current state of the work. Make sure this is the case. Remember how trust works: easy to lose, hard to win.

- Do not complicate! Start from a very simple board, with minimum information and colors. As needs arise, expand this model, but do not hurry. Start by making sure that each addition is truly necessary.

- On the web, look for photos showing Kanban boards. See how they can differ and let yourself be inspired. But do not blindly copy the solutions used by others. They needed them, but does your team need them as well?

- Simple rules are good. Complicated rules are a bane. Kanban is simple. Do not complicate!

- There's no sense to have too many rules regulating Kanban. A handful of simple rules plus common sense is all you need. Say a definitive *NO!* to excessive formalization.

- Imposing solutions is a short-sighted policy. Let the team work out its own rules. The chances that they will be followed will go up.

- Make sure that the team has a solid board, cards, markers and magnets. Make sure that there is a person on the team that will ensure that all the necessary elements are refilled as needed.

- Watch markers and refill their stock! When they are missing, people will start to scrawl with pens, which will greatly reduce the legibility of the information collected on the board.

- Have no mercy for those who scrawl illegibly! …but please start your battle to improve the world with yourself, giving a good example by creating legible cards.

- Take care of the board. Fix smudged lines that separate the individual columns. Move cards slightly to achieve a neater arrangement. Ask the author of an illegible card to recreate it. Verify if the markers write sufficiently well, and, if you decide they don't, make sure there will be new ones. Let the *broken windows theory*[1] inspire you – it says that tolerance for the violation of small rules leads to more serious offences.

- Do not get to worked up on WIP limits. Remember: these are not iron-clad rules, but just pointers meant to encourage discussions and actions that will improve the team's work.

- On the board, also include the tasks that the project owner is expected to work on. By way of example, if a new TV must be bought for the team to be able to admire the CFD chart (see chapter 11.5) on a 60-inch plasma screen, the TV will surely arrive faster if the project owner is asked at every daily meeting on the progress of the work with respect to its purchase.

- The board has a certain capacity. So do human heads. When the team is too large, there will be disorder on the board, and meetings, even the daily ones, which should be short by definition, will get lengthy and ineffective. I do not dare to give an optimal team size, but my experience shows that when the number of people on the team approaches ten, problems begin.

- There will always be a reason why today's meeting can be skipped. Do not give in. Do not disturb your team's rhythm of work.

- When introducing Kanban, remember about the KISS rule, i.e. *Keep It Simple, Stupid!*. Start with something very easy. As weeks pass, you will see what changes really deserve to be introduced.

---

[1] https://en.wikipedia.org/wiki/Broken_windows_theory

Appendix A. Success stories

Kanban turns out to be helpful on very different projects. Let's look at three short descriptions of very different projects on which we successfully used Kanban.

1. A research project. A very unusual project, during which the team had a great deal of leeway on all its aspects, including the choice of the goals to be pursued. It turned out that such freedom of choice can be very difficult. An additional challenge is the fact that external supervision over the progress of the work (and thus external motivation) was practically non-existent. The introduction of the Kanban board was a motivating factor. Since then, we were able to track our progress. In addition, Kanban allowed us to introduce some sort of discipline into our work (for example by defining the DoD), which was important given the limited experience of most team members.

2. Working with an external client. The constant priority changes made it difficult to carry out Scrum iterations. Because we could not depart from Scrum for good, we took as much from Kanban as we could. The board allowed us to see the progress of the work, and it helped propagate knowledge on the current state of work across the team (which numbered 10-12 people).

3. A project for an internal client. As the number of team members increased (from 3 to 6), we have decided to use the Kanban board. This was a great decision. The project owner was the happiest of all, as he finally got the full picture of the current state of work. As we started to use the board, strange cases of the following type also stopped: *"a half-done task that we interrupted for a reason that is difficult to pinpoint, as a result of which our production code has a backend implementation which is not used at all"*. In addition, the level of cooperation among team members also clearly increased.

Appendix B. Personal Kanban

> There are only two real rules with Personal Kanban:
> 1. Visualize your work
> 2. Limit your work-in-progress
> — Jim Benson *Personal Kanban*

Probably, in addition to professional tasks you have many personal projects. Perhaps there is a technology you would like to get to know in depth. Perhaps you want to learn to cook, because you feel you cannot keep going on hot-dogs and scrambled eggs. Maybe you want to do a triathlon, or just jog a kilometer. Perhaps you would like to plan a great vacation for the entire family (and earn the money to pay for it). You surely have many plans and projects, big and small.

Personal Kanban is one of the ways that can facilitate the pursuit of your own projects. Just as in the case of team Kanban, the board plays a central role here. It should include cards representing all the tasks you are working on. And then it is just as in Kanban: you complete a task, you move a card, you take another task.

What do you gain this way? First of all, you will see how many things you are trying to tackle simultaneously. And you will decide not to pursue some of them, making you more likely to succeed in selected ones. Second, the sight of the tasks on the board means that you are focusing on them, not allowing you to spend time on side issues. Third, moving cards to signal that you have completed a subtask on the way to a greater goal is a motivating factor.

I will not write more about this, letting you decide how sensible such an approach is. On the web, you will find a lot of interesting material on personal Kanban. There are some people who are very enthusiastic about this way of acting. Perhaps you will find it valuable as well.

Appendix C. What's next?

> I fear the man of a single book.
>
> — Saint Thomas Aquinas

There are now countless works on Kanban and related techniques. Below, I've collected suggestions on materials that are worth reaching for in order to expand the knowledge gained while reading this book. Needless to say, this is not the end of the journey. Instead, the suggested materials are 'road signs' helping with further progress.

There is a number of nice animated illustrations of the **basic ideas of Kanban** such as WIP (WIP: why limiting work in progress makes sense (Kanban) [https://www.youtube.com/watch?v=W92wG-HW8gg]). There are also interesting films that introduce the 'lean' concepts, e.g. what is lean production [https://www.youtube.com/watch?v=J4v-HjY3R0Y].
It is also a good idea to visit Agile Manifesto [http://agilemanifesto.org].

Make sure to read Matt Sine, The Seven Wastes of Software Development [https://dzone.com/articles/seven-wastes-software]. The idea of *waste reduction* comes from the work of Taiichi Ohno, whom we have already mentioned in this book. It was transferred into the IT world by Mary Poppendieck.

In some way, **Scrum** constitutes competition for Kanban. It is a good idea to get familiar with Scrum; perhaps this is the answer to the problems of your team? Read the Scrum Guides [https://www.scrumguides.org] to find out.

You will find information on the **retrospectives** in practically any guide to Scrum. There is a notable book that is devoted to them in its entirety: Esther Derby, Diana Larsen, "Agile Retrospectives: Making Good Teams Great", Pragmatic Bookshelf, 2006.
Many websites, e.g. Fun Retrospectives [http://www.funretrospectives.com], Agile Retrospective Resource Wiki [http://retrospectivewiki.org] or Retromat [http://plans-for-retrospectives.com] provide detailed descriptions of various exercises used during retrospectives[1].

---

[1]Just please remember that retrospectives are not merely about allowing the team to have some fun…

Tasks to complete, represented in Kanban by cards on the board, encourage taking a closer look at **'user stories'**. You can start here: Wikipedia: User Story [https://en.wikipedia.org/wiki/User_story]. Useful books include Jeff Patton, "User Story Mapping" [https://jpattonassociates.com/user-story-mapping/] and Gojko Adzic, "Fifty Quick Ideas To Improve Your User Stories" [https://gojko.net/books/fifty-quick-ideas-to-improve-your-user-stories/].

If you are interested in personal growth, I encourage you to study materials on **personal Kanban**. You can start the journey on Jim Benson's webpage (Personal Kanban [http://www.personalkanban.com]). From there it is just one step to the topic of productivity, on which so much has been written that it is hard to decide where to start.

Bibliography

Books

Marcus Hammarberg and Joakim Sundén, "Kanban in Action", Manning, 2014

David Allen, "Getting Things Done: The Art of Stress-Free Productivity", Penguin Books, 2002

Jurgen Appelo, "#Workout: Games, Tools & Practices to Engage People, Improve Work, and Delight Clients", 2014

Douglas W. Hubbard, "How to Measure Anything: Finding the Value of Intangibles in Business", 2010

Articles & Blogs

Andy Carmichael, "The difference between Cycle Time and Lead Time... and why not to use Cycle Time in Kanban"

Matt Stine, "The Seven Wastes of Software Development"

Paweł Brodziński, "The Kanban Story: Measuring Lead Time"

Paweł Brodziński, "A Fool With a Tool Is Still a Fool"

Mike Burrows, "Introducing Kanban through its values"

Mattias Skarin, "10 Kanban boards and their context"

Yuval Yeret, "Cumulative Flow Diagrams"

Henrik Kniberg, "One day in Kanban land"

Matthias Marschall, "3 Reasons Why Your Team Needs Rituals"

Hakan Forss, "Habits of Kanban process improvements"

James Betteley "Are *Ready For* Columns on Kanban Boards The Enemy of God?"

Wikipedia, Kanban

Wikipedia, Kanban (development)

Wikipedia, Lean software development

Wikipedia, Little's Law

David J. Anderson, blog

Jim Benson, Personal Kanban

Video

David J. Anderson, Youtube: various Kanban presentations

www.ingramcontent.com/pod-product-compliance
Lightning Source LLC
Chambersburg PA
CBHW070942210326
41520CB00021B/7015